NASA SP-2000-4224

COMPUTERS TAKE FLIGHT: A HISTORY OF NASA'S PIONEERING DIGITAL FLY-BY-WIRE PROJECT

James E. Tomayko

The NASA History Series

National Aeronautics and Space Administration
NASA Office of Policy and Plans
NASA History Office
Washington, D.C.   2000

Library of Congress Cataloging-in-Publication Data

Tomayko, J.E. (James E.), 1949-
    Computers take flight: a history of NASA's pioneering digital fly-by-wire project/
    James E. Tomayko.
      p. cm.— (NASA history series)
    "NASA SP-4224."
    Includes bibliographical references and index.
        1. Fly-by-wire control—Research—United States—History. 2. Crusader (Jet fighter plane)—History. 3. NASA Dryden Flight Research Center—Research—History. I. Title.
  II. Series

TL678.5 .T65 2000
629.135'5—dc21

                                    99-047421

Dedication

To the women with whom I share my life, for sharing me with NASA: my wife Laura and my daughter Gabriela Huiming

Contents

Acknowledgments ii

Foreword... iv

Preface vi

Introduction: The Promise of a New Flight-Control Technology vii

Chapter One..1
    The History of Flight-Control Technology

Chapter Two 21
    The Origins of NASA's Involvement in Fly-By-Wire Research

Chapter Three 35
    The Reliability Challenge and Software Development

Chapter Four 57
    Converting the F-8 to Digital Fly-By-Wire

Chapter Five 69
    The Phase I Flight-Research Program: Digital Control Proven

Chapter Six 85
    Phase Shifting: Digital Redundancy and Space Shuttle Support

Chapter Seven...103
    The Phase II Flight-Research Program: Proof of Concept, Space Shuttle Support, and Advanced Experiments

Chapter Eight..125
    The Impact and Legacy of NASA's Digital Fly-By-Wire Project

Appendix: DFBW F-8C Flight Logs 136

Glossary..154

Bibliography 161

About the Author 169

Index 170

The NASA History Series 176

Acknowledgments

There are many in NASA I want to thank, most especially Kenneth Cox, John D. (Dill) Hunley, and Kenneth J. Szalai. While I was researching the Space Shuttle control system, I interviewed Ken Cox, who was the first to tell me about the F-8 digital fly-by-wire project. He inspired me to visit the Dryden Flight Research Center to find out more about it, a visit that resulted in a short paper about the project and the desire to do more. Over a decade later, Dill Hunley convinced Ken Szalai that a history of the project was in order. Szalai, then the Center Director, supported the idea and Dill husbanded the project at every stage. He has been a constant source of encouragement and help, truly facilitating the production of this book. He is the finest editor with whom I have ever worked.

I would also like to thank the Dryden archivists who were so helpful in finding source material and answering questions, Curt Asher and Peter Merlin. Pete is a walking library and museum in himself. Betty Love, a veteran of many years at the Center and now a part-time volunteer archivist and researcher, can identify just about everyone who worked there beginning in the Muroc days down to the present, and was very supportive on my research visits. She, amazingly, found the names of all but one of the people in the photograph shown opposite the title page despite 27 years of distance and no identifications kept with the photo.

Ken assembled a team of participants to review each chapter for technical and historical accuracy. I interviewed all of them as well, so they made a significant contribution of their time to this book. They are Dwain Deets, Philip Felleman, Calvin Jarvis, James Phelps, and Ken Szalai. Gary Krier also checked large parts of the manuscript and was very encouraging. These men did a good job of keeping me out of trouble, but I am still responsible for the content of the book and its accuracy. The survival of a particular individual's notes often seemed purely serendipitous. I regret that I did not have the resources to locate a wider variety of diary-like material, but the notes we did find were from key players in every step of the project.

One of the advantages of being a college teacher is the opportunity to hire bright young people to help with projects. Hopefully they learn something about writing history and the subject technology, and they often provide insights someone too close to the material will miss. Debbie Martin built a general bibliography of fly-by-wire that I have used in writing several articles and this book. Megan Barke enthusiastically learned enough about the technology to ask questions that led to improved explanations and was a great help on research trips to Draper Laboratory and the Center. She reviewed and helped revise the introduction and chapters one through four. She also drafted Appendix I and one-third of Appendix II. Alina Mason reviewed and helped revise all remaining chapters. She had a keen eye for undefined acronyms and identified items for the glossary as well as doing the last two-thirds of Appendix II. Mackenzie Dilts helped me through a few emergencies and revised the bibliography and appendices. Rachel Knapp worked on the index. To all I express my gratitude and best wishes for success in their lives.

On the production side, Carla Thomas and Jim Ross scanned the illustrations—a tedious job that required considerable concentration, expertise, and effort on their parts. Camilla McArthur shepherded the book through printing under a Government Printing Office contract and Darlene Lister did the copy editing, both in expert fashion. Steve Lighthill made this a book by doing the layout. I am grateful for a special version of Paul McDonell's digital image of the CF-105, of which others are available at http://www.comnet.ca/~mach3gfx.

I would like to thank my colleague Mary Shaw for encouraging me to apply to do this project and for her moral support. Finally, my wife, Laura Lallement Tomayko, helped gather some materials from the archives on a research trip, and also helped with the revisions and footnotes, including the entire bibliography.

James E. Tomayko
13 August 1999

Foreword

This history of the F-8 Digital Fly-By-Wire Project at NASA's Dryden Flight Research Center by Dr. James E. Tomayko of Carnegie Mellon University is important for a number of reasons. Not the least of these is the significance of the program itself. In 1972 the F-8C aircraft used in the program became the first digital fly-by-wire aircraft to operate without a mechanical backup system. This fact was important in giving industry the confidence to develop its own digital systems, since flown on military aircraft such as the F-18, F-16, F-117, B-2, and F-22, as well as commercial airliners like the Boeing 777. Flying without a mechanical back-up system was also important in ensuring that the researchers at Dryden were working on the right problems.

Today, digital fly-by-wire systems are integral to the operation of a great many aircraft. These systems provide numerous advantages over older mechanical arrangements. By replacing cables, linkages, push rods, pull rods, pulleys, and the like with electronic systems, digital fly-by-wire reduces weight, volume, the number of failure modes, friction, and maintenance. It also enables designers to develop and pilots to fly radical new configurations that would be impossible without the digital technology. Digital fly-by-wire aircraft can exhibit more precise and better maneuver control, greater combat survivability, and, for commercial airliners, a smoother ride.

The F-8 Digital Fly-By-Wire Project made two significant contributions to the new technology: (1) a solid design base of techniques that work and those that do not, and (2) credible evidence of good flying qualities and the ability of such a system to tolerate real faults and to continue operation without degradation. The narrative of this study captures the intensity of the program in successfully resolving the numerous design challenges and management problems that were encountered. This, in turn, laid the groundwork for leading, not only the U.S., but to a great extent the entire world's aeronautics community into the new era of digital fly-by-wire flight controls. The book also captures the essence of what NASA is chartered to do— develop and transfer major technologies that will keep the U.S. in a world leadership role as the major supplier of commercial aviation, military, and aerospace vehicles and products. The F-8 project is an example of how advanced technology developed in support of the agency's space program, in this case the Apollo endeavor, can be successfully transferred to also address the agency's aeronautics research and development goals, greatly multiplying payoff on taxpayer investments and resources. It is truly an example of what NASA does best.

Dr. Tomayko tells this story very effectively, which is another reason his history is important. For the first time, he makes the details of the development of digital fly-by-wire available in one place. Moreover, he does so in a style that is both readable and accessible to the general reader. He brings to the task unique qualifications. Besides being trained in the history of technology and a seasoned author in that field, he also has over 20 years of experience in the computing industry and academia, where he has taught and published about real-time systems and software engineering. This combination of talents and experiences has allowed him to tell an important story exceptionally well. I commend his book to anyone interested in the history of technology and/or aviation.

Calvin R. Jarvis
Former Director, Dryden Aerospace Projects
30 August 1999

Preface

One hundred years after the Wright brothers' first powered flight, airplane designers are unshackled from the constraints that they lived with for the first seven decades of flight because of the emergence of digital fly-by-wire (DFBW) technology.

New designers seek incredible maneuverability, survivability, efficiency, or special performance through configurations which rely on a DFBW system for stability and controllability. DFBW systems have contributed to major advances in human space flight, advanced fighters and bombers, and safe, modern civil transportation.

The story of digital fly-by-wire is a story of people, of successes, and of overcoming enormous obstacles and problems. The fundamental concept is relatively simple, but the realization of the concept in hardware and software safe enough for human use confronted the NASA-industry team with enormous challenges. But the team was victorious, and Dr. Tomayko tells the story extremely well.

The F-8 DFBW program, and the technology it spawned, was an outgrowth of the Apollo program and of the genius of the Charles Stark Draper Laboratory staff. The DFBW program was the high point of my own career, and it was one of the most difficult undertakings of the NASA Dryden Flight Research Center. It was not easy to do the first time in the F-8 and it will not be easy to do in the next new airplane. I hope the history of this program is helpful to the designers of the DFBW systems that will enable new and wonderful aerospace vehicles of the future.

Kenneth J. Szalai
F-8 DFBW Principal Investigator
Former Director, NASA Dryden Flight Research Center
5 October 1999

Introduction:
The Promise of a New Flight Control Technology

The 25th of May 1972 was a typical day at Edwards Air Force Base in the high desert of California: sky brutally blue, light winds—perfect flying weather. In the 25 years since Chuck Yeager had broken the sound barrier there in the X-1, the Edwards main runway and the surrounding dry lakebeds had seen more than their share of exotic flying machines. Some, like the X-1 and the X-15, were highly successful, while others failed miserably. Today it was not an X-plane that taxied slowly to the end of runway 04, but a rather conventional-looking F-8C fighter whose appearance belied its internal modifications. In fact, it was the first of the F-8C designation, built in 1958. Despite its unassuming appearance and vintage, this airplane would usher in a new era in aviation—the era in which flight control would revolve around a digital computer, and the pilot's inputs made through stick and rudder would be a small part of a flood of data from sensors and switches that enable the computer to stabilize an unstable airplane. Successful tests with this conventional airframe would pave the way for the unconventional: the Space Shuttle Orbiter, the B-2 flying wing, and commercial airliners like the Airbus A-320 or Boeing B-777 with smaller, lighter control surfaces and almost unimaginable reliability.

Research pilot Gary Krier climbs into the cockpit of the F-8. He eventually flew 64 missions, more than any other pilot in the project. (NASA photo E-24682).

For NASA research pilot Gary Krier, this was all far in the future, though not as far as the engineers on the digital fly-by-wire project thought. He was more interested in the next hour of flying. Krier had significant flight experience, including flying chase for the X-15. He called that job a "rite of passage," meaning he had "arrived" as a member of the small fraternity of test pilots. A slim, brown-haired 38-year-old who looked at least 10 years younger, Krier was facing his first research program as project pilot. Krier knew that thorough preparation was the best foundation for a safe flight. He had "flown" over 200 hours in the "Iron Bird" simulator that replicated the hardware and software in a modified F-8 airframe sitting in a hangar. Krier also flew practice missions in an unmodified F-8, NASA 816, convincing himself that if he had engine power and rudder authority he could land on one of the many lakebed runways scattered throughout restricted area R-2515. He also knew that there were *two* fly-by-wire systems on board, in case one of them failed. The primary was centered on an Apollo Guidance Computer of the type used on the lunar lander. The secondary was a three-channel redundant electronic analog computer system like those pioneered by the German engineers who built the A-4 (V-2) in World War II.

What prompted NASA's engineers to put a digital computer in an airplane? They hoped it would be the next victory in the aeronautical designers' perpetual war against size and weight, while gaining considerable advantages in maneuverability and safety. An aircraft has only two fundamental tugs-of-war among four forces to deal with: thrust versus drag and lift versus weight. The NASA project concentrated on the second conflict. By actively controlling an aircraft's surfaces automatically, it is possible to stabilize it artificially, thus reducing the need for horizontal and vertical stabilizers with large surface areas. In military flying, this means higher thrust-to-weight ratios and greater weapons loads; in commercial flying it results in more paying passengers.

At around 10:00 a.m., Krier taxied onto runway 04. Even though the prevailing wind usually favors runway 22, the opposite is used for first flights since it has thousands of feet of empty lakebed at the end of its length of 15,000 feet; enough room for a straight-ahead emergency landing if something strange happened on takeoff. Despite the presence of the backup flight control computer, no one wanted to test it for the first time close to the ground. At 10:14, Krier applied full power and the F-8C, painted white with blue lightning bolts on its sides, launched itself, its pilot, and the world into the digital era of aviation.

NASA's digital fly-by-wire project is remarkable for its impact on the evolution of flight control systems but is also a case study in how engineers sold ideas and conducted research at the NASA Flight Research Center (redesignated the Dryden Flight Research Center in 1976) during the late 1960s and throughout the 1970s. Arguably, the fly-by-wire project could not be done as easily today since the channels for selling project ideas and obtaining funding are more complex now. This is due to formality and layers of bureaucracy that did not exist in the 1960s. Even in the personnel-bloated Apollo era, NASA engineers knew their headquarters counterparts informally. If a proposed project could get past the Center director and find a champion at headquarters, the engineers and their advocate could usually work out a proposal that would sell, even if the resulting funding was little more than a start. Projects that capitalized on the seed money would, if not prosper, at least survive. As Krier climbed out to the northeast, there was no doubt that the fly-by-wire project had spent its startup money well. Furthermore, the working atmosphere at the Flight Research Center emphasized cooperation and teamwork by all members of a program. Pilots were actively involved in design and planning; crew chiefs, hardware and software people, everyone collaborated to create a successful project. Twenty-five years later, NASA Administrator Daniel Goldin launched his "faster, better, cheaper" campaign and visited all the Centers to explain it in a series of all-hands meetings. He found that the atmosphere he was trying to create at NASA already existed at the Dryden Flight Research Center as the established way of doing business.

Part of the reason that the fly-by-wire project remained cost-effective is that the Dryden engineering team chose to be the system integrators, rather than passing that responsibility to a prime contractor. Aside from saving money, Dryden gained immeasurable experience working with diverse suppliers and computer software. This experience grew throughout the program and was passed on to other projects at the Center.

Another notable feature of the fly-by-wire project was the extraordinary quality of the engineers, pilots, and support personnel who worked on it. Toward the end of the first series of flight tests, pilots from other projects were brought in to check out the airplane. They commented in their flight reports that the project team was performing at a first-class level in all aspects. Many of the key engineers went on to greater responsibility, including Ken Szalai, the lead research engineer who became Center director in 1994. The "F-8 Mafia," as some Dryden employees referred to it, represented a stable group of NASA engineers with over three decades of experience, a situation increasingly rare in the downsized and probably younger NASA of today. They were far removed from an "old boys'" network; they had achieved their positions of leadership through merit.

Finally, aside from carrying out the model "good" project of the 1970s in achieving its own objectives, the team was able to aid the Space Shuttle group in gathering experience with new flight computers and eliminating problems that came up in the Approach and Landing Tests conducted at Dryden. Without the F-8 ready to fly support missions, the Shuttle project would almost certainly have suffered costly and frustrating delays. The F-8 also flew with a prototype of the sidestick planned for use in the F-16. These pieces of technology transfer were only a part of the wide dissemination of results that occurred in papers and industry workshops. The Dryden engineers were thereby able to convey their confidence in fly-by-wire to reluctant commercial airplane builders.

Many technologies with significant advantages fail to catch on due to economic constraints, or sometimes simply because their time has not come. Fly-by-wire demonstrated that its time had come. Within 10 years of Krier's pioneering flight, digital controls were the only technology used in designing the world's military aircraft. The premier warplanes designed in the 1980s, the B-2 flying wing and the F-22 stealthy air-superiority fighter, are both naturally unstable and take advantage of the active control provided by digital fly-by-wire. In a sense, the B-2 is the ultimate demonstration of the conventional cruciform tail reduction that is possible with fly-by-wire: the tail is gone, so there is no passive stabilization. Although not as radical as the B-2, the first commercial aircraft with digital flight controls was also designed within a decade of the initial flight of the modified F-8. The Airbus A-320, which has provided a clinic for cockpit designers through its various crashes, has a high level of success with its control system, which features a unique architecture to enhance reliability. The Airbus flight control system has migrated to all new models of that firm's aircraft. Even the conservative industry dominator, Boeing, has launched the B-777 with a digital control system of more conventional design.

The modified F-8 climbs into the future of flight control. (NASA photo ECN-3312).

Participants in the NASA program are uniformly amazed at the rapidity of the transfer of this technology. There were other predecessor and contemporary fly-by-wire programs, and the combined weight of everyone's results no doubt contributed to the widespread use and acceptance of the technology. However, the engineers at the Flight Research Center were the first to use digital computers, and these are the choice of nearly all designers now. They were the first flight-control team to struggle with that troublesome, intractable medium: software. They clearly demonstrated that it was possible to build and integrate a combined hardware and software flight-control system, and that it had all the advantages expected of fly-by-wire with additional flexibility provided by embedding the flight-control laws in computer code. The world was apparently ready, and now the skies are filled with digital computers flying "in very tight formation." This book tells their story.

Chapter One: The History of Flight-Control Technology

When NASA's Flight Research Center entered the field of active control research, it stood at a crossroads reached by previous work. Flight-control philosophies had progressed through several stages of emphasis on inherent and dynamic stability to instability. Flight-control technology initially had limited mechanical means and a great dependence on the pilot. By 1960, control of both unpiloted missiles and automatic control of piloted aircraft greatly reduced this reliance. A technology called "fly-by-wire" — due to its electrical, rather than mechanical, nature — made these accomplishments possible. Fly-by-wire was not fully born of itself. At first, it was the solution to a set of control problems — a fix, not considered an earth-shaking discovery. Later, as its true potential was revealed, researchers at NASA, the Air Force, and elsewhere began to build flight-test programs around fly-by-wire in order to exploit it more effectively. NASA, however, took a road different from the others. This book examines the history of control technology and early fly-by-wire to understand why.

The Flight-Control Problem

When you consider the relatively simple technology that went into the Wright Flyer, you might wonder why it took so long to achieve powered flight. A number of inventors and researchers were very close to beating the Wrights into the history books. The stumbling block facing the brothers and their competitors was the inability to maintain even straight and level flight without extreme effort. At a meeting in Chicago of the Western Society of Engineers in September of 1901, Wilbur Wright summarized the situation: "Inability to balance and steer still confronts the students of the flying problem.... When this one feature has been worked out, the age of flying machines will have arrived, for all other difficulties are of minor importance."[1] Ironically, the Wrights would "solve" the problem by reliance on the skills of the pilot. Seventy years later, the computer would move in for humans as the cornerstone of the solution.

The Essence of "the Flying Problem"

Until the early nineteenth century, fledgling aeronauts thought the solution to what Wilbur Wright termed "the flying problem" lay in imitating the birds. The Icarus legend and Leonardo da Vinci's stiff-winged, human-powered flying harness are examples of how even the most creative people

---

[1] Wilbur Wright before the Western Society of Engineers, Chicago, 18 Sept. 1901, as quoted in Duane McRuer and Dunstan Graham, "Eighty Years of Flight Control: Triumphs and Pitfalls of the Systems Approach," in *Journal of Guidance and Control*, 4 (July-Aug. 1981): 353-362.

Very unstable in pitch, somewhat unstable in roll, and slightly unstable in yaw, the Wright Flyer took constant intense concentration and great skill to fly. It is no wonder that its longest flight lasted a mere 59 seconds. (Photo courtesy of the Smithsonian Institution).

centered on a natural dead end. The interplay of forces seemed impossible to figure out. Weight, drag, and thrust were easy enough, but overcoming weight to achieve lift seemed to be a matter of exerting sufficient downward thrust—impossible to achieve by a human or early mechanical systems.

Nevertheless, things "flew": pieces of paper, kites, and other rigid or semi-rigid objects dependent on random gusts of wind. Finally, in the early years of the 1800s, the Englishman George Cayley figured out that a rigid plane moving through the air generates lift, and the world changed. For the remainder of the century the problem shifted to the need to provide sufficient airflow over planar "wings" to generate enough lift to balance the weight of the aircraft. Even a flat surface gives the lifting effect if sufficient forward speed is applied.[2] As some wags used to say about stocky jet fighters, "Even a brick can fly if you hang a big enough engine on it."

As work progressed, researchers found out that making the wing flat on the bottom and curved on the top helped generate more lift with less thrust. This is the shape of airplane wings today. The differences in the thickness of the wing are a function of the forward thrust: more thrust, thinner wings. This is why jets have thinner wings than piston-driven airplanes.

Sir Hiram Maxim, probably better known as the inventor of an effective machine gun, used the profits from his invention to study flight. He built an enormous aircraft that actually lifted a few inches off its long launch track. The emphasis of his research was directed at the power plant (perhaps

---

[2] Charles Stark Draper, "Flight Control," 43rd Wilbur Wright Memorial Lecture, *Journal of the Royal Aeronautical Society*, 59 (July 1955): 451-478.

naturally, given his mechanical talents). Other experimenters, such as Otto Lillienthal, Percy Pilcher, Octave Chanute, Samuel Pierpont Langley, and Charles Manly, concentrated on improving lifting surfaces. Their gliders achieved high degrees of inherent stability, or the tendency to maintain straight and level flight given constant acceleration. They evidently thought that this was a desirable trait, perhaps driven by the fact that their early models were unpiloted, and the intended demonstration was of steady lift.[3]

The problem with inherent stability is that it has a double-edged nature. It causes the aircraft to respond quite strongly to gusts, making it difficult for the human pilot to control the response. Glider pilots of the era had to balance and steer by shifting body weight. Ironically, both unstable and stable designs required the athletic ability of a gymnast to achieve the same effect. One of the key difficulties was achieving static longitudinal stability, or the ability to stay balanced fore and aft. This form of stability is least in low-winged monoplanes at low speeds, which pretty much characterizes the situation of most of the gliders.[4] Maxim experimented with biplane designs. He reasoned that they essentially doubled the leading edge of the wing without doubling the physical width of the airframe. Biplane designs also reap some stability gains.[5]

The difficulty of achieving stability frustrated the aeronauts to the point where Wright made the statement quoted above, and leading aeronautical theoretician G. H. Bryan intoned: "The problem of artificial flight is hardly likely to be solved until the conditions of longitudinal stability of an aeroplane system have been reduced to a matter of pure mathematical calculation."[6] Bryan made this particular prediction in June of 1903, and it was published on 7 January 1904, precisely three weeks after the Wright brothers "solved" the "flying problem" once and for all.

The Wright Solution

It is difficult to imagine a more unstable vehicle at low speeds than the two-wheeled bicycle. Even so, most of us cannot remember what it was like to learn to ride one, once the integration of balance, steering, and forward speed is accomplished and practiced. When the bicycle-building Wright brothers turned their restless inventiveness toward the "flying problem," they initially thought that the other aeronauts were past the training-wheel stage, and they worried that they were behind.[7] There existed equations of lift and notions of

---

[3] Charles Stark Draper, "Flight Control," pp. 455, 460.
[4] Melvin Gough, "Notes on Stability from the Pilot's Standpoint," *Journal of the Aeronautical Sciences*, 6, no. 10 (Aug. 1939): 396.
[5] Hiram S. Maxim, *Artificial and Natural Flight* (London: Whittaker, 1909), p. 100.
[6] G.H. Bryan and W.E. Williams, "The Longitudinal Stability of Aerial Gliders," *Proceedings of the Royal Society of London*, 73 (1904): 100.
[7] Orville Wright, *How We Invented the Airplane*, Fred C. Kelly, ed. (New York: David MacKay, 1953) is a

stability, even the concepts of wing warping and vertical and horizontal stabilizers.

When the Wrights tried to apply their competitors' previous work to their own gliders, they found enormous knowledge gaps. They used models and full-size gliders, both piloted and unpiloted, and even had to develop a wind tunnel to settle airfoil design questions. This step-by-step exploration took several years, each one punctuated by an extended field trip to Kill Devil Hills in North Carolina to ride the steady winds available there.

At first, the Wrights tried the stability solution, following the lead of their immediate precursors.[8] As they gained more experience, they began to build their gliders with less stability, depending on the pilot to compensate. Later, Charles Stark Draper, the engineer who led the development of the Apollo lunar spacecraft control system, would point to this decision as the key contribution of the Wrights to aeronautics. They believed in the concept of a stable system made up of machine and pilot as opposed to simply a stable airframe.[9] The center of the problem thus shifted from determining and maintaining some form of inherent stability to that of allowing dynamic stability within controllable limits. Likewise, the emphasis of their work changed from stability to control, opening a door through which computer technology would later walk.

Research done after the Wrights achieved flight clearly showed that oscillations along the longitudinal axis of an aircraft are more easily damped by pilots than are lateral oscillations.[10] It is therefore not surprising that the Wrights solved the longitudinal control aspect first. They added a lever that moved the horizontal stabilizer located in the front of the pilot, who lay on the lower wing. The large surface area of this bi-planed stabilizer — and the fact that both of the small wings moved together — made it fairly powerful as a control surface.

The lateral stability problem, especially in turning flight, took longer to solve. First the inventors tried to use wing warping alone, placing a cradle at the hips of the pilot to control the wires leading to the trailing edges of the wings. After a disastrous flight or two they hit upon the idea of coupling the wing warping to moving the previously fixed vertical stabilizers mounted behind the wings. This scheme enabled coordination of forces resulting in smooth banking turns.

The dependable December breeze was moving at a brisk 27 mile-per-hour clip the day the Wright Flyer sat on the launch rail, ready for its rendezvous with aviation history. Orville Wright lay on the vibrating wing, hips

---

fascinating account of the step-by-step process the Wrights used to solve the stability and control problems.
[8] Draper, "Flight Control," p. 463.
[9] *Ibid.*, p. 461.
[10] Otto Koppen, "Airplane Stability and Control from a Designer's Point of View," *Journal of the Aeronautical Sciences,* 7, no. 4 (Feb. 1940): 137.

centered in the lateral control cradle, thoughts of Wilbur's failure to control the plane three days before clouding his confidence. With the wind at a peak gust, the twin propellers buzzing, and his brother steadying the right wing, Orville accelerated down the single rail gathering speed until the Flyer lifted into the air and flew for some seconds toward a controlled landing in the sands.

The Wrights had made their point: full inherent stability was not a prerequisite for practical flight. This news did not sit well with the other aviation pioneers who labored long on the same path. Maxim seemed least likely to admit to reality. He repeatedly called them "the mysterious Wrights" and stated that "there is much doubt about their alleged flights" even as Wilbur and Orville were steadily piling up time and distance records back home in the gray Ohio skies.[11] As long as fuel tanks quickly ran dry and visual flying was the rule, the lack of inherent stability inhibited no one trying to advance the art of flying.

The Return of the Stability Paradigm

For over twenty years after Kitty Hawk, aircraft of widely varying stability came into common use. Some of the pilot favorites, such as the Spad and the Curtis JN-5 Jenny, were among those unstable in one axis or another. The development of aeronautical engineering was still at the craftsman stage. Bryan's and others' attempts to derive the mathematical equations of stability and lift met with mixed success. Bryan himself reasoned that people did not study mathematical descriptions of stability simply because of the actual success of machines without inherent stability.[12]

It is interesting to review the state of the airplane design art as it stood in about the middle of World War I. F.S. Barnwell produced a slim volume that led the reader step-by-step through the design of an airplane, with an extended appendix by W. H. Sayers that summarized the understanding of stability at that time.[13] Sayers makes it clear that inherent stability as a concept was not perfectly understood. The terminology used to describe it was not standard, or even unambiguous. For instance, within the range of "stable" aircraft were those that were "livelier" due to the concentration of weight near their centers and those that were "steadier" due to distribution of weight. In the former, small forces had a greater effect than in the latter. The range from "livelier" to "steadier" is not defined.

During the 1920s aeronautical engineers began to gear their designs back toward the idea of inherent stability. The reasons are obvious to any automo-

---

[11] Maxim, *Artificial and Natural Flight*, p. 109.
[12] Bryan and Williams, "Longitudinal Stability," p. 3.
[13] W.H. Sayers, "A Simple Explanation of Inherent Stability," in F.S. Barnwell, *Airplane Design* (New York: Robert M. Mcbride and Co., 1917), pp. 73-102.

bile driver on a midwestern interstate: long stretches without outside stimulation, such as distance flying at night, tend to increase fatigue. As ranges of aircraft increased, the need to make them easier to fly also increased. The inability to let go of the controls for even a few seconds works against these objectives. Also, the more stable an aircraft, the easier it is to build a practical autopilot.

There was a debate early in this century between what Charles H. Gibbs-Smith calls the "chauffeurs" and the "airmen."[14] For the former, airplanes ought only to be steered, much like an automobile. For the latter, they must be piloted, like the Wrights flew theirs. After some years of dominance by the airmen, the less talented and more practical wanted a chance at flying for the business of moving cargo and passengers—tasks better fitted to sedate aircraft with "cruise-ship" handling characteristics. The result was a procession of frankly overbuilt machines with large stabilization and control surfaces. However, new demands and the potential for remarkable performance improvements started the pendulum swinging back the other way.

The Benefits of Abandoning Inherent Stability

With an Me 109 on your tail spitting lead at a thousand rounds a second, you cease to be interested in how easy it is to fly a particular airplane straight and level and become radically more concerned about how easy it is to maneuver it in a rapid and unpredictable manner. The simple fact is that if the aircraft is too stable, it is more difficult to maneuver in certain desirable ways. It was recognized as early as Barnwell's 1917 manual of airplane design that "too much inherent stability should not be given to an airplane."[15]

Research studies made just prior to World War II bore this out. Engineers realized that "The idea that the easiest airplane to fly is one that will fly itself was proven false many years ago."[16] They discovered that pilots preferred aircraft with some lateral instability. Even modern light planes flown by recreational pilots do not return to complete wings-level flight after being disturbed by a wind gust.[17] To achieve that level of stability requires increases in the size of vertical stabilizers, and thus more weight and drag.[18] Basically, "the greater the stability, the more demands placed on the pilot and the rougher the flight."[19]

The understanding developed that the real need was for a degree of dynamic, rather than inherent, stability — that is, that the aircraft return to

---

[14] Charles H. Gibbs-Smith, *Aviation: An Historical Survey from Its Origins to the End of World War II* (London: Her Majesty's Stationery Office, 1970), p. 58.
[15] F.S. Barnwell, *Airplane Design* (New York: Robert M. Mcbride and Co., 1917), p. 63.
[16] Koppen, "Airplane Stability and Control," p. 137.
[17] *Ibid.*, p. 135.
[18] *Ibid.*, p. 139.
[19] Gough, "Notes on Stability," p. 395.

straight and level flight after a disturbance in a "reasonable" period of time. The oscillation (disturbance) in question here is referred to as the "phugoid" motion, and it has been the subject of considerable experiment and analysis ever since.[20]

However, the small degree of instability designed into most airplanes would not help much in evading the Messerschmitts of World War II, and certainly not the F-15s and their missiles in the present world. Also, as World War II came to an end, the familiar airplane silhouette of wings, tail, and propeller encountered a revolution as propless turbojets laced the sky over Germany, quickly followed by tailless delta-winged craft. As aircraft performance saw sudden improvement (cruising speeds doubled in less than a decade), new problems of stability, control, and maneuverability came to the fore.

Also, being able to fly into the transonic region (on either side of the speed of sound) caused even more control headaches because of the rearward shift in the center of lift.[21] In the postwar period, researchers and designers realized that the solution to the control problems of supersonic flight and the demands for maneuverability in new aircraft lay in building planes that were less stable, or even unstable in one axis or another. It was one thing to settle on a design and degree of passive stability when the operational speed range was 100 knots, and another when it was 1,000.

In addition to making it easier to design aircraft to operate over a wider performance range, abandoning inherent stability made it possible to reduce size and weight, those twin obstacles to even greater performance. All aircraft depend on balancing weight with lift and drag with thrust. The heavier the weight, the greater the lift needed, and thus, with thrust held constant, a larger wing size is required. That means even more weight and drag. Similarly, the more drag there was from "useless" appendages (such as external weapons stores or cargo pods), the greater the need for more thrust. By allowing instability in one axis or another, the size of the horizontal and vertical stabilizers could be reduced, saving both size and weight. This allowed a performance gain with no increase in thrust and lift.[22] The savings achieved by relaxing the stability requirements can be quite remarkable. Boeing studied converting its KC-135 Stratotanker to a relaxed stability aircraft and estimated a 25 percent decrease in airframe weight with no loss in payload capacity.[23]

---

[20] For example, see William F. Milliken, "Progress in Dynamic Stability and Control Research," *Journal of the Aeronautical Sciences*, 14, no. 9 (Sept. 1947): 493-519, a report from the Cornell Aeronautical Laboratory of extensive experiments with a B-25J and various autopilots to obtain the derivatives of dynamic stability.
[21] Fred Reed, "The Electric Jet," in *Air and Space*, 1 (Dec. 1986-Jan. 1987): 44-45.
[22] Major J.P. Sutherland, "Introduction to Fly-By-Wire," *Proceedings of the Fly-By-Wire Flight Control System Conference* (Dayton, OH: Air Force Flight Dynamics Laboratory, Air Force Systems Command, 1969), p. 260.
[23] J. Morisset, "Fly-By-Wire Controls are on the Way," in *Telonde* (Dec. 1983): 8.

The airplane in Figure 1a demonstrates the situation in an inherently stable aircraft.[24] The center of gravity is forward of the center of lift, so the aircraft would tend to pitch nose-down if it were not for airflow pushing downward on the horizontal stabilizer. The negative result is that this configuration requires relatively large horizontal surfaces, with the accompanying increases in size and drag.

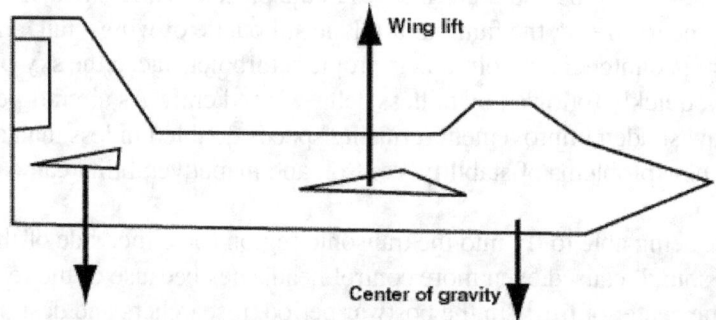

Line drawings 1a, 1b, and 1c originally created by the author and converted to electronic format by the Dryden Graphics Office.

Figure 1b shows the case of a longitudinally unstable aircraft. Here the center of gravity is behind the center of lift. The horizontal stabilizer is now free to produce positive lift, allowing not only a smaller tailplane, but also smaller wings, because all the horizontal surfaces are now overcoming the weight of the aircraft. Moving the center of gravity aft actually increases the range of an aircraft, other things being equal, because it allows more efficient use of thrust. This is especially true when going supersonic.[25]

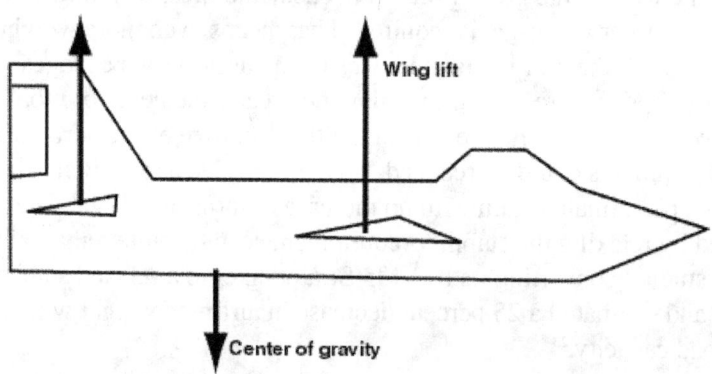

Figure 1c has its horizontal-stabilizing surface in front of the wing, the

---

[24] This and the following two figures are based on D.C. Anderson and R.L. Berger, "Maneuver Load Control and Relaxed Static Stability Applied to a Contemporary Fighter Aircraft," (AIAA Paper 72-87), p. 2.

[25] Ed Daley, et al., "Unstable Jaguar Proves Active Controls for EFA," in *Aerospace America* (May 1985): 34.

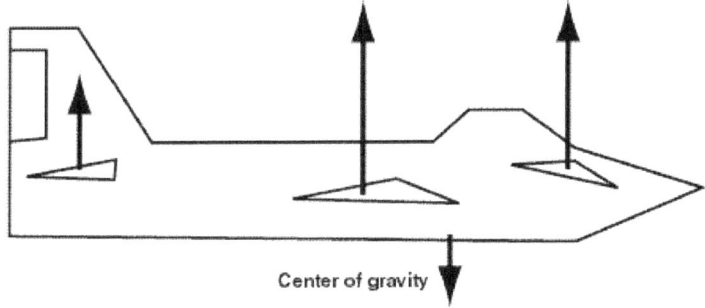
Center of gravity

canard configuration.[26] In this case the aerodynamic center of the aircraft is moved forward due to the lifting surface of the canard. This makes the center of gravity virtually move aft, increasing maneuverability without physically moving the center of gravity.[27] There are many maneuverability advantages to the use of canards, including making it very difficult to stall the aircraft. However, the resultant configuration is also inherently unstable. To use the advantages of new technology and inherent instability, some form of active control must be provided, and only computers were fast enough to help.

The Concept of Active Control

As we have seen, the trend in airplane design was from less stable, pilot-controlled aircraft, toward more stable, less pilot-work-intensive aircraft. The decision to adopt unstable designs in order to increase performance, however, could not simply occur with a greater pilot workload. In fact, with greater performance, the aircraft exceeded the capability of human pilots to control them without some mechanical aid. It is one thing to bring a Jenny moving at 70 knots in a light wind back to straight and level flight by pilot control, but quite another to stabilize a jet fighter at 1,000 knots, or a large aircraft with plenty of inertia. There is simply not enough time for a human being to react. Therefore, the trend reversed from emphasizing stability to emphasizing control.

Initially, control augmentation took the form of extending the pilot's powers in some way. As an example, most light general-aviation aircraft use simple cable systems to connect control surfaces to the control yoke. The pilot physically moves the surfaces using human strength. Place the same pilot in a large commercial aircraft with control surfaces weighing thousands of pounds, moving in a 500-knot slipstream, and even a powerlifter would have trouble moving the control surfaces through direct cable connections. The solution is to install hydraulic systems much like the power steering in a large automobile. Other types of these systems were used to maintain level

---

[26] R.B. Jenny, F.M. Krachmalnik, and S.A. Lafavor, "Air Superiority with Control Configured Fighters," *AIAA Journal of Aircraft* (May 1972): 372.
[27] *Ibid.*, p. 373.

flight. For instance, some high-performance jets such as the Boeing B-47 bomber needed dampers in one axis or another to maintain stability since even small design discrepancies translated into big control problems.

These systems form the basis of what was to come in the fully computerized era, but in themselves they could not do the job. For instance, in a variable-geometry aircraft such as the F-111, F-14, or B-1, the pilots would have a different "feel" as the wings changed the shape of the overall aircraft. These aircraft have "stability augmentation systems" that essentially make the pilot-aircraft interaction feel the same no matter what the position of the wings. The development of these systems forced the inclusion of sophisticated feedback mechanisms and paved the way for the next innovation: the actively controlled airplane.

In essence, the true advantages of instability can only be realized if the pilot is not aware that his airplane is unstable. This means that some method of ensuring stability without constant pilot attention must be used. The solution lies in active control systems that are made possible by the application of computers to monitor and create the necessary feedback. Not only do these fly-by-wire systems compensate for instability, they open up a frontier of previously unthought-of improvements in handling, safety, and utility. Aside from reducing weight, as we saw above, the application of active control has had other positive results. For instance, engines can be smaller due to the reduced weight and drag of the airframes, even without sacrificing maneuverability.[28] The weight and balance problem caused by having both internal and external weapons stores is solved because the control system automatically compensates for differences in flying characteristics as the center of gravity shifts.[29]

However, the most valuable benefit of using active control is that aircraft with highly desirable characteristics such as stealth would simply be impossible to fly otherwise. The Lockheed F-117A, which has proved in combat to be an effective fighter-bomber, is known among its pilots as a smooth flyer, despite the inherent instabilities of its airframe and the multiplicity of radar-deflective surfaces that also deflect smooth airflow over the fuselage.[30] However, new technologies of active control, like sensors, actuators, and computers, needed to evolve before such planes could be built.

Fly-by-wire became possible due to the convergence of existing flight-control technology and specific research and development of sensors, effectors, and computers. Researchers knew they would need certain components to make active control systems. NASA and the Air Force then pursued these "enabling technologies." The result was a unifying system that has made

---

[28] *Ibid.*, p. 376.
[29] G.H. Hunt, "The Evolution of Fly-By-Wire Control Techniques in the U.K.," in *11th International Council of the Aeronautical Sciences Congress* (10-16 Sept. 1978), p. 69.
[30] Draper, "Flight Control," p. 468; Bill Sweetman and James Goodall, *Lockheed F-117A: Operation and Development of the Stealth Fighter* (Osceola, WI: Motorbasics, International, 1990), p. 76.

such innovations as side-stick controllers, high-g cockpits, and integrated engine and flight controls practical.[31]

Control augmentation began as early as the 1890s with Hiram Maxim's experiments with his rail-launched airplane. He devised a gyro device to steady the control surfaces, maintaining longitudinal stability.[32] In the 1920s and 1930s, Elmer Sperry and others pioneered the design and use of autopilots. The seemingly primitive mechanical versions of these reached an apex in the late 1940s when a C-54 aircraft flew from Newfoundland to England entirely under the control of a flight program on punched cards.[33] The amazed pilots were limited to "tending the store," much as they do in this day of flight management systems and digital autopilots.

As high-performance aircraft became the norm, control augmentation based on electrical analog devices advanced the technology of active control. The Anglo-French Concorde supersonic airliner is an example of the application of active control in the transonic flight regime. The next steps, to fully fly-by-wire systems and then to the integrated flight and engine controls of current aircraft, were relatively short. Furthermore, they benefited from the application of systems built for space flight. The development of this technology has caused an evolution for the role of the pilot. The pilots have generally accepted their new role as "systems manager." The practical application of digital computers in the heart of these integrated navigation, stability, and thrust management networks has helped them in their transition.[34]

Active Control in History

The key difference between mechanical control systems and fly-by-wire systems is that the former are distance dependent and the latter are force dependent.[35] A pilot pulling back on the control wheel of a light plane deflects the elevators upward. The distance the elevators move is proportional to the distance the control wheel is pulled away from the instrument panel. The actual proportions are a function of the cable length connecting the control wheel and control surface.

In a fly-by-wire system, the control device is usually a stick, either in the normal between-the-legs position or on the side armrest. Depending on the type of sensor used, either the distance of deflection of the control stick or the force applied by the pilot's hand is measured. This measurement is what

---

[31] Robert L. Kisslinger and Robert C. Lorenzetti, "The Fly-By-Wire Systems Approach to Aircraft Flying Qualities," in *NAECON* (15-17 May 1972): 205.
[32] Maxim, *Artificial and Natural Flight*, pp. 92-93.
[33] Duane McRuer and Dunstan Graham, "Eighty Years of Flight Control: Triumphs and Pitfalls of the Systems Approach," in *Journal of Guidance and Control*, 4 (July-Aug. 1981): 357.
[34] Robert Bernhard, "All Digital Jets on the Horizon," in *IEEE Spectrum* (October 1980): 38; John H. Watson "Fly-By-Wire Flight Control System Design Considerations for the F-16 Fighter Aircraft" (AIAA Paper 76-1915), AIAA Conference, San Diego, CA., 16-18 Aug. 1976, p. 6.
[35] Reed, "The Electric Jet," p. 43.

is communicated to the computer system as the pilot's desires. This makes it possible to control the motion of an aircraft, rather than the surface positions of the elevators, rudders, and ailerons.[36] The result is a new way of piloting. The transition from the old control-centered style to the new is best illustrated by some examples from history. The examples take us from children's vehicles to the moon.

Bicycles

When Charles Stark Draper gave the annual Wright Lecture to the Royal Aeronautical Society in 1955, he chose the title "Flight Control" and spent a considerable portion of the talk discussing the Wrights' decision to allow some instability in their flying machine.[37] Afterwards, A. R. Collar of the University of Bristol commented to the audience, "The Wright brothers, before they became interested in aviation, were manufacturers of bicycles, and of all the unstable machines, I do not know of one more so than the bicycle." So, how is it possible to control one?

The bicycle is limited to motion in two-dimensional space. In steady state, with thrust applied through pedaling, the vehicle moves along on two wheels with little difficulty. However, the beginning rider quickly learns that the initial startup is the most unstable phase of a bicycle trip. Before cruising speed is attained, the rider struggles with the shift of body weight to each side of the spinning wheels and the uneven thrusts of legs on pedals.

On a bike the rider is the sensor suite, central computer, and actuator for the bicycle control system. By shifting body weight, applying thrust, and steering, the rider eventually overcomes the instability of the bicycle and creates a dynamically stable moving system. Even though it is easier to

Controlling an unstable vehicle like a bicycle quickly becomes second nature. (Personal photograph of James Tomayko).

---

[36] Kisslinger and Lorenzetti, "Fly-By-Wire Systems Approach," p. 206.
[37] Draper, "Flight Control."

balance the bicycle on its wheels when in steady motion, the rider still has to apply control forces for maneuvers such as turns. Not only must the front wheel be turned in the desired direction, but the weight of the rider must also be shifted to balance centrifugal forces. No matter how unstable the bicycle is, few potential riders fail to advance beyond the training wheel stage, though many have scrapes and contusions as evidence of difficult lessons learned.

It was this very method of weight shifting and control manipulation that made the flights of late nineteenth-century gliders and the early twentieth-century Wright powered airplanes possible. The Wrights learned to fly much like a child learns to ride a bike; the picture of one brother racing alongside the wing of the flyer balancing it for the other brother during the build up to takeoff speed is reminiscent of a parent running with a young rider, offering instructions and encouragement. Eventually the minds of men turned to the construction of flying machines without pilots, and the technology that led to fly-by-wire control began to develop.

The German A-4 Rocket (V-2)

The interwar years in Germany saw a great upsurge of glider flying. Shackled by the Versailles Treaty, which restricted the construction of powered aircraft, the government encouraged gliding with the intention of building a strong base of pilots for the future Luftwaffe. A young student at the Technical University of Darmstadt, Helmut Hoelzer, was caught up in this fad.[38]

One day while soaring he thought about the limitations of his instruments. In his mind he began to fashion an electrical network that would result in the instantaneous calculation of the true velocity of his glider; that is, the velocity without wind effects. Thinking that this would be a great topic for one of his academic requirements, Hoelzer approached an assistant professor named Kurt Debus (ironically, later the director of the Kennedy Space Center) and asked for some small funding to buy parts. Debus turned him down, citing lack of money and the "tremendous difficulty" of the task.

Hoelzer reluctantly put the project on the back burner. Within a few years, Hitler started World War II, and the now-graduated engineer found himself drafted and sent to the Army's top-secret Peenemünde research base in the Baltic. Wernher von Braun assembled a team there to build the first large ballistic missile, the A-4, later renamed by the German hierarchy as the infamous V-2. Assigned to the guidance system development section, Hoelzer discovered that there were severe problems in controlling the big rocket. The liquid propellants moved around in their tanks, the rocket had to

---

[38] The author obtained information about Hoelzer's life and his work in a series of interviews in the summer and fall of 1983. For a more detailed description of the origins of analog computing and its application to simulation and control of the A-4, see James E. Tomayko, "Helmut Hoelzer's Fully Electronic Analog Computer," *Annals of the History of Computing*, 7, no. 3 (July 1985): 227-240.

pitch over at a specific instant or it would not hit its aiming point, wind gusts affected it, and so on. Earlier rocket programs faced these issues, but not on the scale that the A-4 demanded.

Von Braun's team decided to use graphite vanes mounted in the exhaust of the engine to turn the rocket much as a rudder in the water turns a speeding boat. These vanes were coupled to tabs on the tips of the stabilizing fins, which increased the corrective forces while traveling through the atmosphere. The control system had to keep the rocket balanced, maneuver it at the correct points, and keep it on course.

In the early versions of the A-4, a directional radio beam similar to the British "Oboe" and German "Egon" blind bombing systems indicated the flight path the rocket needed to take. The rocket followed the invisible radio beam pointed at the target. The guidance and control group devised a variety of simulators to make certain that this complex system worked. These ranged from the expensive and full-scale to laboratory versions. An example of the former is the structure that supported a full-size A-4 suspended on a gimbal. The rocket actually ran its engine and the control system tried to keep it

A German A4 (V-2) is readied for a test flight. (Photo courtesy of the Smithsonian Institution).

righted. This sort of test depended on a lot of components working correctly at one time.

The problems mounted when the live-fire tests ended in catastrophic engine failure before the control system could be tested, or if the control system itself ruined the test in some way. The engineers on the project thus sought a way of isolating the work on the control system. Initially they used a mechanical simulator, which turned out to be unsatisfactory.[39] Hoelzer designed an electronic version that was much more effective. Many of the components in this simulator modeled those that the A-4 needed for lateral guidance and attitude control. Hoelzer combined the beam rider circuitry with the attitude control system to form what he called the *Mischgerät*, or "Mixing Computer." This was the first fully electronic active control system and was the basis for the use of analog computers in aircraft flight control systems as advanced as those in the F-16 and F-117A. The A-4 contained the essential components of a fly-by-wire system: sensors, a central computer, and navigation information. Operation Paperclip transferred this technology directly to the United States after the war. Soon thereafter, the von Braun team led the development of a series of rockets of ever-increasing size. These eventually resulted in the capability to orbit artificial satellites as well as fly to the moon.

The Avro CF-105 Arrow

The use of fly-by-wire technology in the A-4 exemplifies the technology as a solution to a control problem, a solution put in place to control existing hardware. In the 1950s, Avro Canada also applied it as a solution to aircraft control. Due to the vast territory of Northern Canada and the threat of Soviet nuclear bombers coming over the Pole, the Canadian Air Force issued a request for proposal for a Mach-2-plus interceptor that could execute a 2g turn at 50,000 feet without losing altitude. In addition, it had to deliver a large payload of air-to-air missiles under ground control. These specifications exceeded the capabilities of any fighter, in service or in planning, worldwide.

Avro proposed a large aircraft (about as long and tall as the Lancaster bombers it had built in World War II) with a high-mounted delta wing, coke-bottle "area-rule" fuselage, and a skyscraper vertical tail. The initial intention was to use a mechanical control system and couple the remote automatic interception interface to the autopilot.[40] However, the designers discovered a

---

[39] Otto Hirschler, telephone interview, 1983. Unless otherwise noted, all interviews cited in this book were conducted by the author.

[40] During World War II, a Junkers engineer named Fritz Haber designed a system to provide control over a composite aircraft called the *Mistel*. It was a run-out Ju 88 filled with explosives, with an Me 109 or FW 190 fighter mounted on top to guide it to its target. Thus, the entire weapon was a three-engined biplane on takeoff and until the attack. Potentiometers were installed in the fighter so that control stick motions would be electrically signaled to the bomber's autopilot. Haber thus claims the first aircraft fly-by-wire system. Once again, it was a solution to a flight control problem.

A digital artist's rendition of one of the CF-105s in flight. (Courtesy of Paul McDonell).

yaw instability that would have required a much larger vertical stabilizer. Their solution was to adopt a dual-channel fly-by-wire system made of analog circuits. At first, the system had no "feel," since control surface motion was a result of stick motion generating an electrical signal, not moving a cable. Eventually Avro had to install springs to provide imitation forces for the pilots. This aircraft is obscure because it had its official rollout on 4 October 1957, the same day the Soviets launched the first Sputnik. Nevertheless, when it had its first flight on 25 March 1958, fly-by-wire had come to high-performance aircraft, if only to provide a yaw damper extended to three-axis flight control.

The Canadian government canceled the CF-105 in 1959. The beginning of the next decade marked a shift in the use of fly-by-wire. Rather than focusing on fly-by-wire as a solution to immediate problems, aeronautical designers began exploiting the technology for its own advantages.[41]

The Apollo Lunar Module

The epitome of active control, prior to fly-by-wire in aircraft, was the Lunar Module of the Apollo spacecraft. Even though the A4 had an active control system, it also had passive assistance during the most difficult parts of its trajectory: early ascent and final descent. This assistance came from the large fins attached to the base of the rocket. As the A4 climbed through the lower atmosphere and passed the "max-q" [dynamic pressure] transonic

---

[41] The Arrowheads [pseudonym], *Avro Arrow* (Toronto, Canada: Stoddart Publishing Co. Limited, 1992).

region it gained significant stability simply because the fins were moving through air.

Even though it is now quite practicable to design a rocket control system that does not depend on aerodynamic assistance, relatively few are built that way. Generally, the fins are used, except in cases where they are more trouble than they are worth, such as in silo-launched intercontinental ballistic missiles. In the case of missiles intended for atmospheric flight, such as air-to-air weapons, the fins are one of the effectors of the control system. However, the Lunar Module could not depend on aerodynamic assistance in any form. It was the first piloted vehicle designed to operate throughout its entire flight envelope in an airless environment. As such, it was necessary to provide the craft with all the components later needed for fly-by-wire aircraft.

Prior to the Space Shuttle, and even now on all the Russian piloted space flights, the crews ride in what are essentially ballistic reentry bodies with little lift. During ascent, the crew faces forward, toward the aerodynamically tapered nose. During descent, they face to the rear of the direction of flight, as the blunt end of the spacecraft is used to slow it during entry into the atmosphere. The heat generated by friction with the air is dissipated by using materials that burn off the spacecraft, which means that this type of capsule is reusable only with extensive repair. No United States spacecraft of this type has ever been reused.

The early spacecraft carrying human beings shared one common characteristic: they lacked any wings or fins. This was entirely due to the desire for simplicity. At the same time McDonnell-Douglas was building the Mercury and Gemini capsule-type spacecraft for NASA, Boeing was working on the X-20 Dyna-Soar project for the Air Force. The X-20 was to be a piloted stub-winged craft mounted atop a Titan booster. Had the project continued, the X-20 would have had the distinction of being the first fly-by-wire aerospacecraft. However, the project was canceled, leaving the field to the initially simpler Mercury and Gemini, and the eventually more complex Apollo.

The control system in the Mercury space capsule consisted solely of an attitude controller designed to work in free fall. The guidance system on the booster rocket controlled the ascent portion of a Mercury suborbital or orbital mission. In the case of the Redstone, it was an electromechanical guidance system using technology not much improved from that of the A-4. The Atlas booster used a ground-based computer to calculate steering signals based on inputs from radar stations.

Once separated from the booster, the Mercury spacecraft was in a ballistic path existing in a relative vacuum. Small jets could be used to induce pitch, roll, and yaw movement. Similar reaction-control jets had already flown on the North American X-15 winged aerospacecraft to control it when it rose far enough above the atmosphere that its conventional control surfaces were not much use. In Mercury, these jets were connected to a fly-by-wire system with a mechanical backup.

The logical design consisted simply of a control signal that transmitted "on/off" commands for the firing of the jets. The attitude of the spacecraft could be changed by the pilot's moving a hand controller, with the direction of the controller's movement indicating pitch, roll, or yaw to the control system. The control system then sent the appropriate signals to fire the correct set of jets to achieve the desired effect.

The Mercury spacecraft could only perform one maneuver that was not strictly attitude control: entry into the atmosphere. For that, the pilot used the control system to point the blunt end of the spacecraft, with its rocket pack, in the direction of flight. When the rockets on the pack were fired, a reduction of velocity occurred that was sufficient to drop the spacecraft out of orbit. The attitude control system then kept the spacecraft's heat shield aimed along the axis of descent. All of this could be done automatically, but the early astronauts, not much removed from their test-pilot days, often insisted on "flying" the spacecraft in even the limited ways available.

By the time the Lunar Module was ready to fly, the pilot was fully integrated into the spacecraft. Surveyor spacecraft void of human passengers made successful automatic soft landings on the moon several times in the mid-1960s. However, the Lunar Module was a much larger and heavier craft, and had the additional restriction of carrying two humans considerably less sturdy than the instrumentation on Surveyor. When Apollo 9 flew into earth orbit in early 1969, the Lunar Module and Command Module practiced for the first time the dance they would do before the otherworldly audience of the Moon's craters.

The flight control mechanism on the Lunar Module was the Primary Guidance, Navigation, and Control System, which was referred to as PGNCS, pronounced "pings."[42] MIT's Instrumentation Laboratory (later Charles Stark Draper Laboratory) developed the system for NASA. Engineers had gained considerable experience by developing the guidance and control for the Polaris submarine-launched ballistic missile and the backup guidance system for the intercontinental Atlas during the late 1950s. They also used some Air Force independent research and development funding to plan a reconnaissance mission to Mars with the objective of taking one photograph and returning it to Earth. They worked on the Mars probe in 1958, and thinking about the interplanetary navigation problem served the team well when they eventually received the NASA contract for Apollo's guidance and control three years later. They had proposed having an onboard digital computer in the probe to make maneuver calculations, and this experience resulted in several concepts eventually incorporated into the PGNCS.

The Apollo guidance and navigation system had all the elements of later

---

[42] James E. Tomayko, *Computers in Spaceflight* (Washington, DC: NASA CR-182505, March 1998), Chapter Three.

systems used in aircraft, but its operational scenarios differed in one major respect from most atmospheric vehicles: it spent long periods in coasting flight. Once the spacecraft entered an orbit around the Earth, the Moon, or in the highly elongated elliptical trajectory between the two, it needed no further assistance from the control system until just before the next powered maneuver. On the lander, PNGCS functioned only in the descent to the moon and ascent into lunar orbit to rendezvous with the waiting Command Module. This means that there was a requirement to realign the sensors prior to any engine firing. The crew accomplished this task as part of the pre-maneuver preparations.

The PNGCS used inertial measurement units in three axes to sense accelerations. Therefore, the units had to have one axis lined up with the centerline of the thrust vector from the engines. Once the crew accomplished this, the inertial measurement units would generate analog signals proportional to the accelerations sensed as the engines fired. Converted to digital data words, the information entered the flight computer software as parameters for the powered flight-control routines. The software compared the changes in velocity in all axes to the pre-calculated target velocities; it then issued commands to main engines and attitude-control thrusters.

When the Command Module executed a maneuver, the pilots could often assume a "hands-off" position, given that the most frequent reason to use the main engine was various orbit changes. However, the PGNCS on the Lunar Module had to allow for frequent pilot input, especially during descent to the surface. This is why the PGNCS software on the lander was in some ways more sophisticated than its brother in the Command Module.

The first lunar landing had all the drama of any science fiction film, yet most of what really happened was unknown to the world at large at the time. As Neil Armstrong and Buzz Aldrin descended to the surface, the computer received so many requests for interrupts from the sensors that it began to get behind in its critical processing. This situation resulted in restarts that caused both crew and ground controllers considerable concern, so much so that there nearly was an abort. With scant seconds of fuel remaining, Armstrong had to fly the Eagle away from a boulder field that was in the primary target area. In overcoming these obstacles, the fly-by-wire PGNCS proved its capability.

The difficulty and romance of lunar flight inspired many projects across NASA centers. The Flight Research Center was no exception. Its indigenous Lunar Landing Research Vehicle—a project to explore techniques for simulating lunar landings—involved people and ideas that would carry over into the fly-by-wire program.

Chapter Two: The Origins of NASA's Involvement in Fly-by-Wire Research

The idea of lunar flight even caught the imagination of engineers outside of NASA's space centers. At the Flight Research Center at Edwards Air Force Base, a small group designed and built the Lunar Landing Research Vehicle, or LLRV, in conjunction with Bell Aircraft. This project had no direct organizational connection to Apollo. Its objective was to explore solutions to flying in the environment near the lunar surface and to simulating that environment.[1] Work began in 1961, with some hope of influencing the final Lunar Module design.

The Lunar Landing Research Vehicle flying at the Flight Research Center in 1967. (NASA photo ECN-1606).

The Moon has one-sixth the Earth's gravity, so the major problem in building the LLRV was negating the remaining five-sixths. The solution was a vertically mounted jet engine that had enough thrust to support the vehicle's weight in such a way as to give the pilot the feeling he was operating in lunar gravity. The LLRV was not aerodynamic. It used reaction jets for maneuvering. The Flight Research Center chose a type of fly-by-wire system to control it.

By the late 1960s, Flight Research Center engineers had several examples of NASA's use of fly-by-wire: the LLRV, the piloted spacecraft, the X-15 reaction control system, as well as the lifting bodies. Also, the X-15

---

[1] Neil Armstrong, personal letter to author, 8 Mar. 1998; interview of Hubert Drake by J. D. Hunley, Ames Research Center, 16 Apr. 1997, p. 4.

was winding down, the LLRV was over, and it was time to move on to different work. The Flight Research Center aims to serve the aviation community by experimenting with new technologies, so the engineers started brainstorming projects that would have high impact. A group led by Melvin E. Burke and including Calvin R. Jarvis, Dwain A. Deets, and Kenneth J. Szalai, thought that a project to explore the capabilities of fly-by-wire in aircraft would have the most leverage.[2] They reasoned that commercial airplane manufacturers would be hesitant to adopt the technology without extensive evidence that it worked. Also, they wanted to explore the concept of active control and its capabilities to revolutionize airplane design. They felt that the time was right to do this work, as the enabling technologies had reached a stage where they could be used to build active control systems. Burke, a hardware specialist, knew that what such a project needed was highly reliable sensor, computer, and actuator components. By the end of the 1960s, all of these had evolved enough to be useful in a fly-by-wire airplane.

Maturation of the Enabling Technologies

In an active control system centered on a computer, sensors provide input and the actuators execute the output commands. This is a feedback system in which control-surface deflections caused by the actuators change the state of the sensors, which affects the output from the computer, and so on. Such systems had already been used in autopilots and rocket guidance as well as in the LLRV. The sophistication of these sensor and feedback systems rapidly increased in the 1940s and 1950s as research into flying qualities led to the development of "variable-stability" aircraft. The Cornell Aeronautical Laboratory built a series of these special planes, which consisted of existing airframes with equipment added to change their handling characteristics. There are two types of such variable-stability systems: "response feedback" and "model following." In the former, the scheme is set up to sense the aircraft response to a particular pilot input and then feed that to the control system. In the latter, onboard computers "force" the aircraft to respond in a way identical to the model of the target aircraft.[3]

One of the most long-lived and famous of the variable-stability test aircraft was a T-33 that had the long nose of an F-94 attached to it in 1954 to hold the control system.[4] This aircraft evolved over a 35-year period from relatively primitive analog controls to advanced digital systems. Besides flying-qualities research, the modified T-33 was used to mimic other aircraft to work out control problems. One of its last projects was simulating the

---

[2] Melvin E. Burke, telephone interview, 17 February 1998; Dwain Deets, interview, Dryden Flight Research Center, 5 Jan. 1998.
[3] Waldeman O. Breuhaus, "The Variable Stability Airplane from a Historical Perspective," unpublished manuscript, 26 Feb. 1990, p. 5.
[4] Breuhaus, "Variable Stability Airplane," p. 30.

Swedish Saab JAS-39 Gripen fighter aircraft. Its fly-by-wire flight control system was the suspected cause of a crash, and the variable stability aircraft helped verify changes in the software. This technology has been adapted to other more permanent uses such as the Gulfstream jets using Sperry flight-control equipment modified as landing trainers for the Space Shuttle orbiter. One place in the two-pilot cockpit closely resembles the Shuttle's cockpit. Aside from routine landing training at the Johnson Space Center and the Dryden Flight Research Center, an aircraft is often ferried to potential Shuttle landing sites during missions. In case winds are questionable, an experienced astronaut can fly landing approaches in the Gulfstream to determine how the Shuttle itself would handle.

Aside from the variable-stability aircraft's valuable contributions to flight control capability, other more limited flight-test programs helped contribute to the maturity of the enabling technologies. In the late 1960s, a B-52 flew with an analog-based flight-control system activated from the left pilot seat. It explored the potential for "structural mode control," such as overcoming the loss of a major portion of the tail or wing.[5] Other operational aircraft, such as the F-111 and Concorde, had stability augmentation. The electrical component of fly-by-wire systems showed up in the French Mirage fighter series as early as 1963.[6]

So parts were nearly in place. It only remained for NASA to assemble them in an existing aircraft to prove the principle and lay the groundwork for operational use of the fly-by-wire concept. This was easier said than done, however, since the three enabling technologies came from such different roots and were not necessarily compatible.

Sensors

Sensors are carried on all aircraft. Depending on the sophistication of the autopilot and navigation system, there may be many different types of sensors. As an example, one of the simplest and most prevalent is the pitot-static system, which supplies information to the pilot and flight-control system about airspeed, vertical speed, and altitude. A relatively small hollow tube, the pitot projects from the wing or fuselage of an aircraft in such a way that there is an unobstructed flow of air into it. The pressure of this ram air is compared with the pressure of stable outside air gathered through a port mounted away from turbulence. Through comparison of the two pressures, indicated airspeed can be calculated.

In the case of the pitot-static system, the "sensors" are completely passive: the pitot tube and static port essentially sample the air directly. Also, since the static air sample is taken without any correction or correlation to the

---

[5] R. C. Ettinger, "The Implications of Current Flight Control System Integration," in *Society of Experimental Test Pilots*, 15, no. 2 (24-27 Sept. 1980): 19.

[6] J. Morisset, "Fly-By-Wire Controls are on the Way," in *Telonde* (Dec. 1983): 8.

movement of the air outside the aircraft, the airspeed indicated on a gauge is not the ground speed. This is because it has no way of allowing for the effects of wind. An aircraft with a direct 20 knot-per-hour tailwind would actually be moving relative to the ground at the indicated airspeed plus 20 knots. This means that the pilot is responsible for calculating the effects of wind using weather data and computers that contain information about the overall impact of wind from all directions.

Gyroscopic instruments are more complex. A gyroscope tends to resist forces applied to it once it is spinning. Thus, an instrument such as an attitude indicator can use the position of a gyroscope to correctly show changes in the angle of the wings and nose of an aircraft relative to the horizon. In this case, the sensor is the gyroscope itself coupled with some reference point. In simple aircraft the pilot is responsible for monitoring the attitude indicator to keep the aircraft straight and level or to use it as a reference in turns. In visual flight conditions, the attitude indicator is largely unnecessary since the pilot can use the actual horizon for reference. However, in instrument-based flying, the indicator is crucial to the pilot's ability to maintain orientation.

Gyroscopic sensors can also be used to measure angular velocities. These "rate gyros" are the basis for stability augmentation systems and are important components in fly-by-wire controllers. In the late 1940s, Boeing produced the first swept-wing turbojet-powered bomber, the XB-47 Stratojet. During flight tests an excessive yaw motion occurred at low speed and low altitude, with the problem worsening as wing loading increased (either from extra weight or maneuvering).[7] The solution was the addition of a rate gyro mounted in the yaw axis. The gyro generated a signal voltage proportional to the yaw rate, and that voltage value positioned a push-pull tube to damp yaw motion; this constituted a simple, one-axis stability augmentor.[8]

This technology rapidly proliferated: the British used yaw dampers on the Meteor jet fighter and a pitch damper on a six-engine flying boat. The Northrop YB-49 flying wing also had some stability augmentation added after achieving flight status.[9] Note that all of these stability augmentors came into use as a reaction to problems in actual flight test. They provided valuable experience in the use of sensors and feedback.

Burke used another type of sensor for the fly-by-wire project: an inertial measurement unit.[10] Such devices had been developed in the 1940s. They measure accelerations in each axis of motion. This acceleration data is used in an inertial navigation system to calculate velocity and position without any other sensor input. This is handy in a vacuum. Since good inertial measurement units were common in the space program, they could readily provide

---

[7] Rolland J. White, "Investigation of Lateral Dynamic Stability in the XB-47 Airplane," *Journal of the Aeronautical Sciences*, 17, no. 3 (Mar. 1950): 133.
[8] White, "Investigation of Lateral Dynamic Stability," p. 135.
[9] Breuhaus, "Variable Stability Airplane," pp. 7, 9-10.
[10] Burke interview, 17 Feb. 1998.

data for the computer, the next device in the control chain.

The Role of the Computer

The central component of all fly-by-wire systems is the flight computer. The computer uses control laws specific to an aircraft to calculate the commands necessary to maintain stability and implement pilot desires. Control laws are the equations of motion that have to be solved to actively control an unstable aircraft. The values for these equations are specific to each aircraft design. That is why control laws embodied in electronic analog circuits make those circuits unusable in any other aircraft. There are two types of computers used in fly-by-wire systems: analog and digital. Each type has advantages and disadvantages, and there was considerable debate among the NASA engineers over which to use.[11]

Analog computers exist in a wide variety of forms. In fact, long after the advent of digital computers, there were still many more analog computers in use than the digital ones so familiar today. The log-scale slide rule, once the dominant personal computing device, is an analog computer. It works by creating a mechanical analogy between the positions of numbers on its various scales and the products, quotients, squares, square roots, cube roots, etc., that it is used to calculate. Another type of mechanical analog computer was the differential analyzer, which was in scientific use from the early 1930s through the early 1950s (and was one of the lesser known "stars" of the film *When Worlds Collide*). The "DA," as it was called, had cams of various shapes to model the terms of equations. The analyzer filled a good-sized room and had to be operated by hand.

Such mechanical analog computers are not as practical flight-control devices as their electronic brethren. The German A-4 [V-2] system used such an electronic analog computer. It modeled the differential equations of the control laws and conveniently accepted voltage values as input and generated them as output. These voltages could then be amplified as commands to the actuators of the control system. Thus, by the early 1940s it was possible to use an analog computer in flight control. For nearly forty years thereafter, such devices formed a core enabling technology for fly-by-wire.

The fact that the control laws are hard-wired into an analog computer is both an advantage and a disadvantage. The advantage is that it is difficult or impossible to corrupt an analog computer through power transients, software viruses, or other weaknesses experienced by digital computers. The disadvantage is that to "re-program" an analog computer, one must physically rearrange the circuits into a new structure that models the modified control laws. Furthermore, analog circuits are subject to signal drift in their responses, and this must be compensated, usually by voting of output from

---

[11] Burke interview; Cal Jarvis, interview, Lancaster, CA, 7 Jan. 1998.

multiple circuits.[12] Higher temperatures also affect analog computers because information is in the form of amplitudes, and temperature effects modulate the amplitude. Nevertheless, analog computers were used in the Canadian CF-105 and the first U.S. Air Force fly-by-wire tests. Burke and Jarvis were familiar with them from the LLRV program.

In the late 1950s, when the concept of fly-by-wire first came under serious research scrutiny, the image generated by "computer" was of a multi-ton monster voraciously consuming space and power—hardly an attractive alternative for aircraft control-system designers obsessed with the limitations of size, power, and weight in aerodynamics. Thus researchers only considered digital circuits in limited areas. A 1961 study at the U.S. Air Force Flight Dynamics Laboratory simply replaced analog amplifiers with digital differential analyzers.[13]

Substituting circuits forced the engineers to face the key difference between analog and digital computation. Analog devices depend on a continuous stream of data signals. Digital circuits, by their very nature, need data to be transformed into a stream of bits. The problem is that the signal streams in a complex real-time system might be too dense and rapid for the analog-to-digital converters to deliver all the sensor data to the computer.[14] This means that the data must in effect be sampled, rather than used in totality. The difficulty is in the accurate processing of sampled data in order to make it as useful as a complete data set. It was not until 1963 that the mathematical basis of digital control became widely available due to published work on sampling theory.[15] Note that aircraft systems were not the only beneficiaries of this foundation. Digital control in manufacturing, automobiles, and medical instrumentation has similar problems and has benefited from this information.

Another aspect of digital computers that needed to be improved before they could be used in aircraft was their size. Presper Eckert and John Mauchly had a lot to do with the development of the world's first general-purpose electronic computer, the ENIAC, at the University of Pennsylvania during World War II. It filled a very large room and required significant power and air conditioning to operate, primarily since it used vacuum tube technology. After the war, Eckert and Mauchly started their own computer

---

[12] T.J. Reilly and J.S. Prince, "Relative Merits of Digital and Analog Computation for Fly-By-Wire Flight Control," in J.P. Sutherland, ed., *Proceedings of the Fly-By-Wire Flight Control System Conference* (Dayton, OH: Air Force Flight Dynamics Laboratory, Technical Report AFFDL-TR-69-58, 16-17 Dec. 1969), p. 205.

[13] J.J. Fleck and D.M. Merz, "Research and Feasibility Study to Achieve Reliability in Automatic Flight Control Systems," General Electric Company, TR-61-264, Mar. 1961, p. 80.

[14] G.J. Vetsch, R.J. Landy, and D.B. Schaefer, "Digital Multimode Fly-By-Wire Flight Control System Design and Simulation Evaluation," in *AIAA Digital Avionics Systems Conference* (2-4 Nov. 1977), p. 204.

[15] Jay Roskam, lecturer's notes from "Airplane Stability and Control: Past Present, and Future," Long Island section of AIAA, 16 Mar. 1989, p. 3; Benjamin C. Kuo, *Analysis and Synthesis of Sampled-Data Control Systems* (Englewood Cliffs, NJ: Prentice-Hall, 1963).

company and built a computer for Northrop that was eventually intended to fly in an aircraft. However, as with a graphite-pile nuclear reactor carried in a B-36 in early tests of atomic power for aircraft, no one thought of their computer as a practical device.

The transistor improved the situation tremendously, and a discrete-circuit, transistorized computer built by IBM flew in the Gemini piloted spacecraft in the mid-1960s. It was the development of the integrated circuit that truly made embedded computers in aircraft practical. Early in the 1960s, the Apollo spacecraft development and the Minuteman ICBM (intercontinental ballistic missile) program consumed nearly all U.S. production of integrated circuits for their respective guidance systems. Still, these computers had their logic represented by collections of low-density chips, some, such as the Apollo computer, with as few as four gates. Each gate represented one Boolean function. Current integrated circuits can have millions of gates.

The improvements in digital computer hardware made possible equally important improvements in the capability of the software that embodies the control laws of the aircraft. Whereas with an analog computer the "software" is essentially hardwired into the machine, a digital computer can be adapted to many different uses by changing its programming. A limitation on software for real-time systems in aerospacecraft is the size of a computer word. It not only affects the scale at which the computer can do computations; it affects the flexibility of its instruction set and the application software built for it. Engineers programmed early digital systems exclusively in low-level machine languages that are very difficult to inspect and understand and thus prone to human error. Early recognition of the inherently complex nature of these machine-based languages inspired the development of machine-independent languages such as FORTRAN, which express mathematical formulae in terms more recognizable by the average engineer. However, the use of such high-level languages requires special translation software such as interpreters and compilers that recast the language statements into machine code.

Even though these languages reflected a significant engineering improvement, they were not readily adaptable to the embedded computer systems demanded by fly-by-wire. They lacked statements to support functions such as scheduling of processes. Also, real-time systems have strict performance constraints, and engineering managers thought compiler-generated machine code was too inefficient to meet these requirements.[16]

Effectors and Actuators

The last enabling technology for fly-by-wire flight control consists of the actuators that move the control surfaces. In the mechanically based flight-

---

[16] Reilly and Prince, "Relative Merits," p. 210.

control systems, the control surfaces move under direct-cable positioning. This is replaced by electrical connections to actuators in fly-by-wire systems. In fact, the original meaning of "fly-by-wire" is limited to this technology alone.

Gavin Jenney, one of the pioneers of the technology, working at Wright-Patterson Air Force Base and founder of the aptly named Dynamic Controls, Inc., says that, "When we were developing fly-by-wire, the purpose was to provide safe and reliable electrical control between the pilot and the flight control surfaces as a replacement for the mechanical connection."[17] Such connections did not need either computers or sensors, but rather simple physical force to electrical force converters at one end, and electrically operated hydraulics at the other. Such systems could be made triply or quadruply redundant and still obtain weight savings along with reliability increases over even dual hydromechanical systems. It would have been nearly impossible to achieve practical fly-by-wire without the electrical actuators and their associated equipment.

This is what had been achieved in sensor, computer, and actuator technology when Burke's group was considering fly-by-wire for airplanes. The engineers felt that these technologies had reached a point where they would be practical to use. However, it would be necessary to make hard choices before even trying to sell the program to Center Director Paul Bikle. He has been characterized as "sensitive but not sympathetic." He would listen to an idea, but as soon as it failed to catch his interest, he would simply walk away.[18] They had to have a solid plan before meeting with him. The most difficult choice was whether to use a digital or an analog computer. Most other decisions depended on that one.

Analog versus Digital

There are two ways to send numbers on electrical wires: continuously, or in ones and zeroes. Numbers are transmitted in electronic analog circuits as continuous current at varying voltage levels proportional to the values being transmitted. Volts are a measure of pressure, so the bigger the value, the larger the voltage. Digital signals are sent as streams of bits—binary digits—which can be either ones or zeroes. A specific bit length represents a word of information. Once in the computer, the data is manipulated with the control laws for the airplane. In an analog computer, the equations are represented by circuits that implement the mathematics. In a digital computer, the control laws are in software. This means that analog computers are effectively a single-airplane system; they cannot be moved from one type of aircraft to another without extensive physical changes. In a flight-test program, charac-

---

[17] Personal letter to the author, 18 Mar. 1992.
[18] Gary Krier, interview, Dryden Flight Research Center, 9 Jan. 1998.

terized by continuous tweaking of components, this could potentially be a problem. In a two-phase program like the fly-by-wire project, where an aircraft change is possible between phases, analog computers are even more awkward.

Digital computers are more flexible due to software. The phrase "general purpose computer," which is only applied to digital machines, implies their ability to adapt through different software programs. However, digital computers have advantages in addition to their programmability. By proper scaling of the data represented in digital words, such computers can be made to be more accurate than their analog counterparts. They also can compensate for drift in analog subcomponents. An attraction for the engineers at the Flight Research Center was that with a digital system, they could include some logic in the control laws, making them more robust.[19]

Analysis of the choice between analog and digital computers shows that at the time any comparison made based on considerations of pure size and complexity does not show much difference. For simple systems like short-lived missiles and non-combat aircraft, analog computers are best in most instances. Conversely, most complex systems have long-living applications that benefit from software changes. However, as one flight-control engineer said, "Just where this crossover point lies is difficult to judge." Therefore, the final decision had some political aspects.

This B-47 was modified by the U.S. Air Force as a fly-by-wire testbed using analog computers. (U.S. Air Force photo).

The team at the Flight Research Center initially wanted to go with analog computers.[20] It had experience with them from the LLRV project, plus there was the U.S. Air Force's fly-by-wire test program as a source of experi-

---

[19] D.A. Deets and K.J. Szalai, "Design and Flight Experience with a Digital Fly-By-Wire Control System Using Apollo Guidance System Hardware on F-8 Aircraft," *Proceedings of the AIAA Guidance and Control Conference*, Stanford, CA, (AIAA Paper No. 72-881), 1972, p. 2.

[20] Lane Wallace, interview with Cal Jarvis and Ken Szalai, Dryden Flight Research Center, 30 Aug. 1995.

ence.[21] However, the Air Force had been flying a modified B-47 with fly-by-wire initially in only the pitch axis.[22] Furthermore, it was about to embark on the Survivable Flight Control System project with an F-4 aircraft, which would have a complete three-axis control system based on analog computers. As the group at the Flight Research Center considered this, it decided not to compete with the Air Force and to take the leap into the digital world.[23]

It is not clear that the engineers knew what they were getting into by starting to deal with software. Software's flexibility is a bane as well as an advantage. It is too easy to change and very difficult to change correctly: fifty percent of all software modifications, including defect repairs, result in new defects. By 1972, Dwain Deets and Kenneth Szalai came to think of a digital computer and its software as a patchboard in which any two points could be inadvertently, and invisibly, connected. In an analog system an incorrect connection was more easily visible.[24] Nevertheless, they pressed on with digital technology. The problem then became getting a digital computer suitable for flight control. There were no widely available computers at the time with the size, power requirements, weight, reliability, and performance needed for flight. There were the computers used in piloted spacecraft, however. The first proposal was to use three Gemini spacecraft computers.[25]

Nothing had been settled before Mel Burke and Cal Jarvis went to Washington to find the money for the program. Director Bikle, a sailplane pilot, had the vision to see the aeronautical implications of fly-by-wire and supported the proposal.[26] However, the project had the potential to be a tough sell further up the NASA funding chain. A new project at the Flight Research Center was encouraged to have industry interest in the results. But commercial manufacturers were essentially ignorant of digital fly-by-wire. Moreover, even if knowledgeable, they had to consider the three factors that were essential to any control system choice for commercial aircraft: safety, performance, and cost of ownership.[27] So the selling point for the project

---

[21] James E. Tomayko, "Blind Faith: The United States Air Force and the Development of Fly-By-Wire Technology," *Technology and the Air Force: A Retrospective Assessment*, Jacob Neufeld, George M. Watson, and David Chenoweth, eds. (Washington DC: The United States Air Force, 1997), pp. 163-185.
[22] Neither the Air Force nor NASA's engineers had knowledge of the Canadian analog fly-by-wire CF-105, even though it had first flown 12 years earlier. The Air Force had an exchange officer from Canada working on fly-by-wire at Wright-Patterson Air Force Base, and even he apparently did not realize that the Arrow had been fly-by-wire. One explanation for its obscurity was that the new Labor government in Canada canceled the project, and went so far as to destroy all flying prototypes, tooling, blueprints, and records. Avro responded by firing thousands of employees, with many engineers going on to good jobs in the Apollo program. Canadians today still mourn the loss of the program, since it was arguably the greatest technical achievement of that country.
[23] Calvin Jarvis, interview, Palmdale, CA, 7 Jan. 1998.
[24] Deets and Szalai, "Design and Flight Experience," p. 5.
[25] Jarvis interview, 7 Jan. 1998.
[26] Jarvis interview, 7 Jan. 1998.
[27] J.C. Taylor, "Fly-By-Wire and Redundancy," in *Proceedings of the Fly-By-Wire Flight Control System Conference*, p. 187.

surfaced in the demonstration that fly-by-wire, especially *digital* fly-by-wire, would have sufficient impact to get manufacturers on board.

At NASA Headquarters

Burke and Jarvis had an advantage when they gave their sales pitch in Washington. They had to start at the Office of Advanced Research and Technology, and at that time Neil Armstrong was a Deputy Associate Administrator for Aeronautics. Burke knew Armstrong from the X-15 project.[28] They also had to deal with Peter R. Kurzhals and Frank J. Sullivan, the successive directors of electronic guidance and control research, but Armstrong had an immediate interest that made him their key ally.[29] He wanted to see more technology transfer from the Apollo program. When told of the analog versus digital debate and the difficulty of finding a reliable airborne computer, he said to Burke and Jarvis, "I just went to the moon with one."[30] In fact, the Apollo computer was one of the most reliable ever built (see Chapter Three). With the Apollo program shortened, there were plenty of machines available. Armstrong suggested contacting the Draper Laboratory to explore the feasibility of using modified Apollo hardware and software on the F-8. Burke and Jarvis briefed Dr. George Cherry, head of the Guidance and Control Division at Draper, on the project objectives and he was extremely supportive, beginning a strong relationship between the Flight Research Center and the Laboratory that facilitated the transfer of much space technology to the world of aeronautics.[31] One of the first positive results was the use of Apollo hardware for the first phase of the project. The F-8 team would inherit a solid software development infrastructure and process that would have long-lasting impact on how the Center would build software in the future.

When Burke and Jarvis returned to the Center, an initial budget was in the works. The project was to start in early 1971.[32] For the first year, the allotment was one million dollars, a small amount by space flight standards.[33] The entire project, over a decade long, would cost only $12 million. The major task immediately confronting the engineers was acquiring an inexpensive airplane that could be modified to fly-by-wire.

---

[28] Gary Krier, interview, Dryden Flight Research Center, 9 Jan. 1998.
[29] Jarvis interview, 7 Jan. 1998.
[30] Burke and Jarvis gave this account of their conversation with Armstrong. When asked his memories of the meeting, Armstrong replied that it seemed roughly correct (Armstrong letter, 8 Mar. 1998).
[31] Calvin Jarvis, e-mail to Dill Hunley, 19 Aug. 1998.
[32] Shu W. Gee and Melvin E. Burke, "NASA Flight Research Center Fly-By-Wire Flight-Test [sic] Program," briefing slides and commentary, 1971. Available in the Dryden Flight Research Center History Office.
[33] Jarvis interview, 7 Jan. 1998.

Finding the Testbed Airplane

Even as far back as 1971, a million dollars would not go far toward buying an aircraft, so Burke looked for cheaper, preferably free, alternatives. NASA flew a mini-squadron of Lockheed F-104 Starfighters. Adapting one of those seemed a quick solution. However, discussions with the test pilots and mechanics quickly canceled that. The Center, obviously not using an airplane designed for fly-by-wire, would have to modify the plane for digital control. Some of the more exotic modifications like canards, for instance, could be placed in front of the wing, and the horizontal stabilizer could be removed. The result would be an unstable airplane that would better demonstrate the viability of the concept. The pilots and mechanics pointed out that the most likely location for canards on an F-104 is on opposite sides of the nose. Unfortunately, the F-104 had engine air intakes on either side of the fuselage behind the cockpit. This meant that when the canard surfaces moved, they would disrupt airflow to the engine, with a flameout resulting.[34]

Burke finally contacted Admiral Forrest S. Petersen, who had flown the X-15, and asked for help. Petersen plucked four Chance Vought F-8 Crusaders, the Navy's first supersonic fighter, from their destination at the boneyard and sent them to the Flight Research Center instead.[35] Ken Szalai had a model of an F-8 modified to show the proposed changes that would make it

A redundant electrical actuator of the type that needed to be developed for the F-8. (NASA photo EC71-2942).

---

[34] Krier interview, 9 Jan. 1998.
[35] Burke interview, 17 Feb. 1998.

dependent on fly-by-wire. The horizontal stabilizers were cut off and moved in front of the wing (the F-8's single centerline air scoop was not affected), and the twin ventral fins at the tail were replaced by a single forward-mounted fin. Unfortunately, it quickly became apparent that the F-8's stable configuration, plus sensor additions, computers, and new actuators, would be an engineering and fiscal challenge sufficient to consume all available resources. The futuristic-looking F-8 model resided for over 25 years in Szalai's office, its real-life cousin never built. (Remarkably, the planform almost exactly matches that of the X-31!)

The Split into Phases

As 1970 wound down, planning continued. When Burke left the project for a job at NASA Headquarters, Cal Jarvis took over as project manager. The decision to stay conservative and not change the F-8's aerodynamics was finalized. A less conservative decision was to remove the entire mechanical flight-control system. The Air Force was planning to keep the mechanical system as a backup in the F-4 it was modifying with an analog flight-control system. In fact, on its first flight, it took off using the mechanical system and switched to the electronic while in the air. Jarvis' team thought that would be a bad idea. It would not really force the engineers to face the right fly-by-wire problems.[36] By using a digital primary system and an analog backup, they would be fly-by-wire all the way. The Air Force, for its part, pretty much ignored the NASA program.

It was obvious that a demonstration with a single Apollo computer would not be sufficient for the commercial airplane manufacturers. They knew they would need redundancy, as in all their other systems, for passenger safety. Therefore, the Center engineers decided on a multi-phase program. Phase I would have two goals: ensuring that the technology worked, and developing the tools for moving forward.[37] Phase IB would introduce a two-computer primary system to begin dealing with redundancy. Finally, Phase II would concentrate on gaining knowledge and techniques for highly reliable systems. The project was also planned to move fast. The first flight of Phase I was set for early in 1972, with the Phase II system definition during the mid-1971 to mid-1972 period. The first flight of Phase II was set for the second quarter of 1974.[38] This did not turn out to be the actual schedule, but the objectives did not change. This plan in place, the Flight Research Center engineers worked on setting and achieving reliability goals alongside their software partners at the Charles Stark Draper Laboratory.

---

[36] Wallace interview with Szalai and Jarvis, 30 Aug. 1995.
[37] Deets and Szalai, "Design and Flight Experience."
[38] Shu W. Gee and Melvin E. Burke, "NASA Flight Research Center Fly-By-Wire Flight-Test [sic] Program," briefing slides and commentary, 1971. Available in the Dryden Flight Research Center History Office.

Chapter Three: The Reliability Challenge and Software Development

"I would like to observe that using all the techniques at our disposal...I do not believe we can provide assurance that software of any significant complexity achieves failure rates on the order of 10^{-9} per hour for sustained periods.... Software must generally be buttressed by mechanisms depending on quite different technologies that provide robust forms of diversity. In the case of flight control for commercial aircraft, this probably means that stout cables should connect the control yoke and rudder pedals to the control surfaces."—John Rushby, in a report to the U.S. Federal Aviation Administration, 1993.[1]

This Boeing 777 could not be flown commercially without near-unimaginable reliability in its fly-by-wire control system. (Photo courtesy of the Boeing Co.)

Clearly, the most critical technology in the application of fly-by-wire is a means of assuring reliability. When the NASA Flight Research Center team made the decision to go digital, it entered a world plagued by finicky hardware and untrustworthy software. Reliable systems had been built for spacecraft, but at a cost unlikely to be acceptable to a commercial aircraft manufacturer, and certainly not by Cal Jarvis' minimally-funded project. The F-8 program took advantage of previous research into reliability, especially through the fortuitous choice of an Apollo computer for Phase I and by adopting the entire Apollo hardware and software infrastructure. Even though the Center took on the difficult roles of system integration and human-rating

---

[1] John Rushby, *Formal Methods and the Certification of Critical Systems* (Menlo Park, CA: SRI International, 1993).

for flight, the Charles Stark Draper Lab's development and verification process could be inherited outright. This turned into one of the great cost-saving decisions of the project team—and not only for the F-8 program; later projects involving software benefited as well. None of the engineers so eager to use digital systems had extensive software development experience, so being able to apply an established process helped tremendously. Draper Lab had advanced to the leading edge of reliable hardware and software systems, and the engineers there were excited at the prospect of working on the F-8.[2]

The History of Reliability in Computers

Despite the recent concerns of reliability expert John Rushby, quoted above, digital flight-control systems have demonstrated reliable operation for 35 years, ever since the Gemini program in spacecraft and NASA's fly-by-wire project in aircraft. The aircraft designers using such systems today benefit from the early and pervasive concern on the part of computer engineers about reliability. Early computers, with thousands of fragile vacuum tubes, sometimes operated for only seconds or minutes between hardware failures. Scientists were willing to put up with abysmal reliability because what they accomplished in the few moments of run time far exceeded their manual capabilities over many hours. Also, there were no life-critical, real-time applications. However, the motivation remained to develop reliability technology because it was clear that computers would be around for a long time. Scientists realized that lack of reliability would severely hinder usability across a wide range of potential applications.

It seems unlikely that anyone in the 1940s imagined that future counterparts of their room-sized computers would someday, a quarter-century later, fit into one-cubic-foot boxes and be many times more powerful. The challenge then, as now, was to create confidence that such systems could work as well as mechanical, hydraulic systems over long periods of time. Reliability in such systems is in many ways the sum of the reliability of the parts, but there is no doubt that some parts are considered more worrisome than others. This is particularly true of software. One mantra that came out of the F-8 program concerned the flexibility of digital computers: analog designs must be frozen early in the test program, whereas software could be delivered at nearly any time and also reflect changes in the vehicle suggested by earlier tests.[3] The reliability equation has both hardware and software factors. However, changes in software have a high probability of causing a defect, making exploitation of its flexibility difficult in life-critical systems. But,

---

[2] Philip Felleman interview, Draper Laboratory, Cambridge, MA, 27 May 1998. The Draper Lab was initially called the Instrument Laboratory of the Massachusetts Institute of Technology, as explained below in the narrative.

[3] Kenneth J. Szalai, the lead engineer on the Dryden Flight Research Center's fly-by-wire project, is reported by his colleagues to have noted this.

early in digital computing history, hardware was more of a problem.

Von Neumann's Approach to Reliability and Its Impact on Later Designs

While electronic analog computers were small and powerful enough to work in flight-control systems as early as 1940, no one seriously considered using digital computers for that purpose until the late 1950s. The reason was simple: digital computers were still giants. They used vacuum tubes, they had large power and refrigeration support systems, and their circuitry was neither densely packed nor reliable. As far as aircraft designers were concerned, they were cargo. At that time, the chief cause of computer failure was not software defects, as it is today, but hardware faults. A particular logic circuit could easily lose a vacuum tube and the resulting loss of a bit would result in an error in output. Early computer experts were quite concerned by this, of course, because widespread computer use meant that they would eventually be operated by non-experts who would be much less likely to detect subtle hardware failures.

John von Neumann, one of the true geniuses of the twentieth century, spent much of the last decade of his life thinking about how digital computers are similar to the human brain and about how exploiting the similarities could result in more sensible and reliable machine designs. In January 1952, he gave five lectures at the California Institute of Technology entitled "Probabilistic Logics and the Synthesis of Reliable Organisms from Unreliable Components" in which he suggested a way of increasing the reliability of computer systems.[4] He proposed a component called a "majority organ." This would be used to vote the inputs from redundant strings of logic circuits. He chose as an example a triple-logic design. Von Neumann showed that a positive value e exists for all components that represents their probability of failure. Exploring the mathematics, he noted that eventually all systems still fail, but increasing the number of input bundles to the majority organ allows a designer to fine-tune the desired reliability.

At first glance, this seemed to ensure that digital computers would never find their way into flight-control systems. Objections to redundant digital computers centered on size, power, and weight. Triplicating the logic circuitry and adding majority organs meant a penalty in all three areas. However, within a few years, transistors matured enough to replace vacuum tubes, core memories became more reliable and rugged (though still not very dense), and physical miniaturization together with lower power requirements all became common. Therefore, interest in using digital computers in control systems increased, and von Neumann's elegant proofs became interesting to designers.

---

[4] John von Neumann, "Probabilistic Logics and the Synthesis of Reliable Organisms from Unreliable Components," in *Automata Studies*, C. E. Shannon and J. McCarthy, eds., (Princeton, NJ: Princeton University Press, 1956), pp. 43-98.

By the mid-1950s, engineers in the Flight Dynamics Laboratory at Wright-Patterson Air Force Base had already realized the advantages of fly-by-wire.[5] They began to look at enabling technologies that would make the reliability of digital fly-by-wire more equal to mechanical systems. In 1960, a young engineer named James Morris, who later became one of the primary champions of fly-by-wire at Wright-Patterson, led a project to examine the state of the art in reliability and make suggestions for control system designs. From 1 May 1960 to 30 April 1961, a group of General Electric engineers, led by J. J. Fleck, did literature searches, conducted interviews in the field, and wrote an evaluation of the most prevalent reliability schemes.[6] They found the von Neumann lectures and did projections showing that identical software feeding outputs from individual computers to majority logic voters made failures 300 times less likely in 100 hours of operation.[7] However, some of their conclusions turned out to be untrue and others were ignored. For instance, Fleck and his colleagues wrote that dual-redundant systems would be less powerful than triple-or-greater redundancy due to the need for a resource-grabbing software monitor and hardware switch. Yet the longest-lived spacecraft in operation, the two Voyagers, use three different dual-redundant digital computer systems.[8] Their over-20-year operational history suggests that they are not really impractical. Fleck and his associates also concluded that redundant components with voters are more reliable than redundant general-purpose computers.[9] The controller used for the Saturn V booster later used the voter design. However, the continued shrinkage of computer hardware has made redundancy of entire processor systems more desirable because of their simplicity and interchangeability.

The majority logic voters were quickly identified as candidates for single-point failure that might result in a complete system failure. GE suggested a single-transistor voter of extreme simplicity.[10] Its engineers felt that this type of voter would have reliability equal to that of other hard-wired analog circuits. Eventually, this architecture would find its way into the U. S. Air Force's F-16, its first operational fly-by-wire aircraft. However, the control system used analog computers, with voters. Digital systems rarely use this simplistic method of redundancy management.

Despite these early studies of digital systems, the Air Force at Wright-Patterson chose to focus on flight-control systems using analog computers. As research on using digital computers for aircraft experienced a hiatus in the early 1960s, NASA adopted them for piloted spacecraft. In doing so, it

---

[5] James E. Tomayko, "Blind Faith: The United States Air Force and the Development of Fly-By-Wire Technology," in *Technology and the Air Force* (Washington, DC: U.S. Air Force, 1997), p. 167.
[6] J. J. Fleck and D. M. Merz, "Research and Feasibility Study to Achieve Reliability in Automatic Flight Control Systems," General Electric Company, TR-61-264, Mar. 1961.
[7] Fleck and Merz, "Research and Feasibility Study," p. iv.
[8] James E. Tomayko, *Computers in Spaceflight*, NASA Contractor Report 182505, Mar. 1988, Ch. 6.
[9] Fleck and Merz, "Research and Feasibility Study," pp. 66-67.
[10] Fleck and Merz, "Research and Feasibility Study," p. iv.

pioneered the two methods of achieving the reliability of digital control systems that are still used today.

Redundancy and Backup: the Apollo Experience

For the ten years between von Neumann's lecture and the final specifications of the Apollo spacecraft control system, redundancy, either dual or triple, was the basis for increasing the reliability of fly-by-wire control systems. For the Apollo Program, size, power, and weight drove the search for a single-string system of adequate reliability. The Gemini earth-orbital spacecraft used a single digital computer on flights of up to two weeks duration (although it was not powered up the entire time). A lunar mission was planned to be less than two weeks. Actually, the development of the Gemini computer followed the choice of the Apollo control system designers, and Gemini missions were partly intended as test flights to explore Apollo objectives of long duration, rendezvous, and computer control.[11] The computers for the two programs were quite different.

Later called the Charles Stark Draper Laboratory in honor of its famous founder, the Instrumentation Laboratory of the Massachusetts Institute of Technology had won the contract to develop the Apollo guidance and navigation system. The initial digital computer design was based on the Polaris submarine-launched guidance computer.[12] Draper Lab had to show that the Apollo version of this computer and its resident software could provide sufficient reliability for the lunar mission. The target was 99.8 percent probability that the computer would function at any given instant.[13] Eldon Hall, the chief designer of the computer, quickly calculated that using conventional methods of redundancy to achieve the goal would result in excessive size, power, and weight. Therefore, Hall's group decided to use a single computer, and to test each component at every spot in manufacture.[14] Phil Felleman, who worked on Apollo and later became manager of Draper's F-8 efforts, said that "every piece of metal could be traced to the mine it came from."[15] This decision meant that the reliability of the computer system was essentially purchased through a massive investment of time, money, and energy certifying every part of hardware and software. The gamble paid off: there were 16 computer and 36 display and keyboard system failures in the 42 computers and 64 DSKYs[16] built—all on the ground.[17] With zero inflight failures in 1,400 hours of operation added to the preflight operations, the actual demonstrated

---

[11] Tomayko, *Computers in Spaceflight*, Ch. 1.
[12] Tomayko, *Computers in Spaceflight*, p. 31.
[13] Eldon C. Hall, "Reliability History of the *Apollo* Guidance Computer," Draper Laboratory Report R-713, Jan. 1972, p. 7.
[14] Hall, "Reliability History," p. 12.
[15] Felleman interview, 27 May 1998.
[16] The display and keyboard unit had its name shortened to the acronym DSKY.
[17] Hall, "Reliability History," p. 31-34.

reliability over the life of the program came to 99.9 percent.[18]

Software quickly became the main driver of cost and schedule. The techniques of making reliable hardware were known to engineers before the program began. However, ensuring software reliability was still an immature process. It remained so for many years. Producing code is essentially a lonely act, like writing or making art. Personal styles ranged from composing at the keypunch to meticulous preparation by drawing flow diagrams and thinking through the logic multiple times. Software managers faced the problems of obtaining requirements for the programmers to code and then verifying that the requirements had been implemented. They then had to validate that the resulting system did what it was supposed to do. In projects with teams of programmers, individual coding styles gave uneven results. Therefore, one historical aspect of software development is the struggle to get intuitive programmers to document their management designs, annotate code, and then use repeatable processes, configuration management, and the like.

NASA wanted to use a more standardized development process for Apollo, so it asked Bellcomm, Inc., to do a study of successful software development and management techniques. The resulting report is a good overview of software project management to this day, although the expected

The DSKY eventually used on the F-8 in Phase I of the fly-by-wire program. Warning lights are in the upper left section, displays in the upper right, and the keyboard is in the lower section. (NASA photo EC96-43408-1 by Dennis Taylor).

---

[18] Hall, "Reliability History," p. 40.

proportions of effort for different development phases are somewhat naive.[19] Of the 68 pages in the report, just *one* is devoted to testing. Nonetheless, that page contains the first inkling of the detail required to ensure that the software has a reliability level equal to that of the hardware. The suggestion was to use five layers of verification:[20]

1. Subprogram-unit test, which means that each module of the eventual software is individually tested to find coding defects.
2. Consolidation test, in which the tested units are assembled, piece by piece, into the flight software load.
3. Acceptance test, in which the flight software is verified against the requirements as a whole.
4. Hardware/software integration test, in which the software is tested while resident on flight-equivalent hardware.
5. Live-system environment test, in which the software is executed in the actual conditions it would encounter on a lunar mission. This meant that the true environmental tests were limited to one-G situations (those in which the gravity is equal to that on the earth's surface at sea level) like pre-launch, built-in tests, etc. Tests in the zero gravity of space had to be in a simulated environment.

By October of 1967, the development and verification plans showed significant refinement. This was accomplished in the wake of a tragic fire on 27 January 1967 that killed all three members of the first Apollo crew. That flight, had it taken place, could have been embarrassing for Draper Lab and NASA because the flight software contained known bugs.[21] Several missions were being prepared at the same time, which resulted in a shortage of personnel and verification equipment. Draper Lab had been extremely optimistic when it was awarded the Apollo contract. Its engineers at first expected to build all the software with about eight people, a number that eventually grew to 300 full-time employees.[22]

Howard W. (Bill) Tindall, the NASA manager assigned to oversee the on-board software development, claimed that engineers at Draper Lab began skipping unit tests to save time and resources.[23] It is harder to fix defects

---

[19] W. M. Keese, "Management Procedures in Computer Programming for *Apollo* — Interim Report," Bellcomm, Inc., TR-64-222-1, 30 Nov. 1964.
[20] Keese, "Management Procedures," p. 3.2.4.1.
[21] Frank Hughes interview, Johnson Space Center, Houston, TX, 2 June 1983; Tindall to multiple addresses, "In Which is Described the Apollo Spacecraft Computer Programs Currently Being Developed," memo, March 24, 1967, seen in the former JSC History Office. When exercising the flight software during simulations on the launching pad, the crew discovered discrepancies between the time calculated for firing the engine to return to earth by the onboard system and the ground system; the ground system was correct. There were other errors of this sort.
[22] Felleman interview, 27 May 1998.
[23] Howard W. Tindall, "*Apollo* Spacecraft Computer Programs—Or, a Bucket of Worms," memo, 13 June 1966.

found in integration test than it is in unit test, since it is not easy in the former case to pinpoint the unit responsible for the defect. Therefore, failure to perform unit tests actually costs more time later, in that the bugs easily repaired in unit tests are confounded with others that would only show up in integration or acceptance tests. The Draper Lab software managers knew this, and skipping tests was not to be a common practice. As the final software delivery was pushed back, crew requests for changes were routinely denied. Even known bugs could not be fixed because they were in software stored in the permanent memory of the computer. Most of these would have work-arounds stored in the limited two-thousand-word erasable memory. There were also small programs to back up the primary system stored there.[24] The delay in piloted flight caused by the fire allowed Draper and NASA to clear up the software testing jam and begin work on the Block II system that actually flew to the moon. There were still small glitches in every flight, but nothing serious happened in the flight control and navigation system.[25]

The new development and verification plan of 1967 was intended to define a process that made testing easier, a change in emphasis that would serve the Apollo and later software programmers well.[26] The plan defined milestones used for each flight's software load that had associated tests.[27] These provided synchronization points for both developers and managers:

1. Preliminary design review (PDR) is held after the equations intended for a flight had been defined and verified on an engineering simulator.
2. Critical design review (CDR) is a formal review of design specifications for the software.
3. First article configuration inspection (FACI) is held to approve test plans and results.
4. Customer acceptance readiness review (CARR) is to certify that the software is ready for manufacturing into the permanent and erasable memory loads.

This set of milestones and the overall plan were to be prepared by NASA engineers, with significant help from contractors, especially Draper Lab. It would then be turned over to another contractor, TRW, Inc., for further refinement. By the end of April 1968, it had been split into three plans: software development, verification, and management.[28] The four phases of testing were more clearly defined in the verification plan:[29]

---

[24] Vincent Megna, interview, Draper Laboratory, 27 May 1998.
[25] Megna interview, 27 May 1998.
[26] R. E. Wilson, M. Kayton, W. Gilbert, K. J. Cox, et al., "*Apollo* Guidance Software—Development and Verification Plan," NASA Manned Spacecraft Center, Oct. 1967, p. 3-1.
[27] R.E. Wilson et al., "Apollo Guidance Software," pp. 7-3 to 7-8.
[28] J. V. Mutchler and H. A. Sexton, "*Apollo* CMC/LGC Software Development Plan," TRW Note No. 68-FMT-643, Apr. 1968. TRW stood for Thompson-Ramo-Wooldridge.
[29] Mutchler and Sexton, "*Apollo* CMC/LGC Software," pp. 2-2 and 2-3.

1. Engineering language simulations to validate equations.
2. Bit-by-bit simulation of coding for the guidance computer.
3. Bit-by-bit and hybrid simulations of collections of programs (steering, engine cutoff, etc.).
4. Mission sequencing (using the programs in mission order).

The time between the FACI and the CARR was exclusively devoted to test groups three and four.

This process — with its plans for each area of concern, sharply defined milestones, and continuous verification of the software at each stage of development — is still considered excellent today. It is actually more of a goal than common practice. The Software Engineering Institute of Carnegie Mellon University recently produced a Capability Maturity Model (CMM^{sm}) for software development.[30] The Institute is a federally-funded research and development center with the mission of improving software engineering. The CMM is a benchmark used to assess an organization's ability to produce software. There are five levels of maturity defined in the model. Historically, 15 per cent of these assessments are at Level Three or above. The Apollo software development process of the 1960s is a Level Three, a completely remarkable achievement since the first concerted effort to define and practice "software engineering" did not occur until 1968.[31] The first group (and still one of a handful) to attain a Level Five rating was the Space Shuttle on-board software development team of IBM-Houston, which had inherited the best parts of the Apollo spacecraft software process.

All of the verification and validation would be for naught if the software did not enter the spacecraft in the exact bit-for-bit form that constituted the final version. Since the Apollo computer used permanent memory—woven into ropes of ferrite cores and wire—it was extremely critical that the software in the ropes was correct. The ropes could not be changed without complete remanufacture. Draper Lab had prepared a process for certifying the correctness of rope manufacture as early as the verification plan of May 1965.[32] This process is representative of the care taken in all aspects of development:

1. Code is put on 80-column punched cards.
2. An assembler converts the code to binary on magnetic tape; the tape recording has two-way parity checks with error correction.
3. Simulators use these tapes for testing.

---

[30] M. Paulk et al., "Capability Maturity Model for Software" (Pittsburgh, PA: Software Engineering Institute, Carnegie Mellon University, 1993).
[31] Peter Naur and Brian Randell, "Software Engineering, a Report on a Conference sponsored by the NATO Science Committee, Garmisch, Germany, 7-11 Oct. 1968," (Jan. 1969).
[32] T. J. Lawton and C. A. Muntz, "*Apollo* Guidance and Navigation," MIT, E-1758, May 1965, p. 1-3.

4. A rope-weaving tape and a checker tape on Mylar are created from the tested software and are re-read and checked.
5. Core ropes are made from the weaving tape.
6. The ropes are tested by the checker tape.
7. After installation, the Apollo Guidance Computer self-check program does vertical sum checking of the contents of memory (the parity checks being considered horizontal checking).

The chances of an error escaping on the last test alone are less than one in three billion. This process was very important to the F-8 program, because a planned permanent shutdown of the rope facility meant that there would be only a single opportunity to weave its software.[33]

In the four years between 1964 and 1968, the Draper and NASA engineers had gone from not really knowing the scope of software verification to being able to produce nearly perfect core ropes with only a few non-fatal defects on them. They had also included contractors such as Bellcomm, TRW, and General Electric to gain a wider base of feedback. It was this software process that the fly-by-wire project would inherit.

Despite the invested effort and success of the single-string system on Apollo, it still made some NASA engineers nervous, and the memory and processing deficiencies of the computer added to the discomfort. They wanted some insurance. On 1 August 1964, Joseph Shea, the Apollo spacecraft program director, asked Marshall Space Flight Center to explore using the Saturn V launcher's computer as a backup.[34] The study showed that there was no time in the accelerated schedule to make the modifications, and of course, it could only be used when the spacecraft was still attached to the booster. There was some discussion of having a software program in the Apollo computer containing only those functions absolutely necessary for the return to Earth, lest some undetected error in the main programs arose. However, there was no room for it in the limited memory. Another orphan safety function was the return-to-earth abort from within the sphere of the moon's gravity (about 64,000 kilometers in radius). The Draper team did manage to squeeze in a return-to-earth feature from within the earth's gravity field. In the end, the abort from the lunar sphere was finally handled by pre-computed tables in handbooks, a non-software solution.[35]

Even the most confident engineers did not want to leave one aspect of

---

[33] Felleman interview, 27 May 1998.

[34] Joseph Shea, NASA Apollo Spacecraft Manager, memorandum, 1 August 1964, seen in the former JSC History Office.

[35] General Electric, "Single Failure Point Study of the CSM Guidance and Controls System for the MLL Mission," Contract NASW-410, Task E.4.3.5, Apr. 1968. The report also suggested verification of the lunar module descent engine for use in transearth injection—*Apollo* 13 would later use that engine for course changes during its abort. Ironically, the use of the lunar module as a "lifeboat" on *Apollo* 13 was possible because both it and the command module had identical computers with similar software capabilities for navigation. In effect, a serendipitous redundant system.

the Apollo mission to a single-string system: the lunar landing itself. A computer failure at any point in the lunar module's flight would almost certainly be fatal to the crew. Since redundancy had already been rejected due to limits on size, power, and weight, some form of backup would have to be used. Backup in this context is a system that provides safety but does not provide full functionality. A redundant system would provide full functionality. NASA contracted with TRW to produce a small eight-bit computer called the MARCO 4418 to be the heart of an Abort Guidance System (AGS) for the Lunar Module. Initially, the AGS was intended to provide guidance capability to the crippled spacecraft all the way to a rendezvous with the command module. This requirement was quickly scaled down to simply being able to put the lunar module into orbit. The command module would use its more sophisticated guidance programs to find and retrieve the two astronauts on board. TRW used the same ponderous and effective verification plan as the Draper Laboratory. One report stated that it took an average of nine months to approve, implement, and test a software change, a testament to the inherent size and inertia of such a grand testing scheme.[36]

Despite the general acknowledgment that redundancy would be the most reliable design for a flight-control system, as the 1960s drew to a close only the CF-105 Arrow had used that scheme. Even though the Saturn V was much larger than any previous (or later) booster, its payload weight was so limited that it prohibited redundancy in the Apollo spacecraft.[37] Thus, its designers chose to use a backup scheme. Both the dual-redundant Arrow and the simplex-with-backup Apollo were designed with military pilots or astronauts in mind, people who had many more dangerous concerns than the possibility of computer failure. It was the desire to move this technology into the civilian realm, in commercial aircraft, that forced the adoption of deep redundancy and backup schemes together.

The Reliability Scheme for Phase I of NASA's Digital Fly-by-Wire Project

It was generally acknowledged that the simplex system used in Phase I of the F-8 project would be a proof of concept. The real payoff for commercial manufacturers would come with the redundant system in Phase II. However, even though the mechanical controls were completely removed from the F-8 during its conversion to fly-by-wire, there was never any intention to fly without a backup. The decision to provide safety with an analog system reflected the confidence in the overall concept of fly-by-wire that was demonstrated in other programs.[38] The Air Force's analog fly-by-wire

---

[36] P. M. Kurten, *Apollo Experience Report: Guidance and Control Systems—Lunar Module Abort Guidance System* (Washington, DC: NASA TN D-7990, 1975).
[37] All ground systems for human spaceflight beginning with the Mercury Program (the first) were redundant.
[38] Tomayko, "Blind Faith."

system had a goal of 99.999977 percent reliability during a two-hour mission.[39] NASA eventually wanted to demonstrate 99.9999999 percent reliability with the F-8.[40] In contrast, a speaker at the Air Force's 1968 conference on fly-by-wire technology reported that a triplex stability augmentation system for the SR-71 had achieved 99.95 percent reliability, comparable to Apollo.[41] He revealed calculations that demonstrated a quadruplex system would achieve 99.99065 percent reliability, with FAA records showing commercial aviation historically at 99.999565 percent.[42] Therefore, the F-8, if successful, would be significantly more reliable than commercial aircraft—a worthy goal. The use of dissimilar primary and backup systems in Phase I would achieve progress toward the eventual reliability figure, as the analog backup augmented the Apollo computer's 99.9 percent reliability. Achieving still higher reliability would have to wait until Phase II.

Draper Laboratory Becomes Directly Involved

Soon after Neil Armstrong suggested the use of an Apollo computer on the F-8, the wheels began turning to incorporate the needs of the project into the continuing work at Draper Lab. Because of the Draper Laboratory's unique position in designing the overall Apollo guidance and navigation system, NASA awarded the development of the F-8 digital system under a simple contract process. Philip Felleman was assigned as the first project manager for the Draper Lab. He was trained as a mathematician and had joined the Laboratory in June 1954. Felleman began his career in airborne fire control and then had several jobs during Apollo, including software design and management. He met with Mel Burke, and later Cal Jarvis and Ken Szalai, beginning a relationship with the Flight Research Center based on mutual respect that lasted 15 years. Ken Szalai would later refer to him as "my hero."[43]

The Center set four design rules for both organizations to follow:

1. There would be no mechanical reversion. This forced the use of fly-by-wire completely.
2. There would be no changes to the Apollo system except software— thus, the hardware verification and validation would be inherited intact.

---

[39] David S. Hooker, Robert L. Kisslinger, and George R. Smith, *Survivable Flight Control System Final Report* (Dayton, OH: Air Force Flight Dynamics Laboratory, Air Force Systems Command, 1973).
[40] Kenneth J. Szalai, Vincent A. Megna, *et al.*, *Digital Fly-By-Wire Flight Control Validation Experience* (Report R-1164, Cambridge, MA: The Charles Stark Draper Laboratory, Inc., 1978), p. 113.
[41] J. C. Taylor, "FBW and Redundancy," in J. P. Sutherland, ed., *Proceedings of the Fly-by-Wire Flight Control System Conference*, (AFFDL-TR-69-58, Dayton, OH: U.S. Air Force, 1969), p. 190.
[42] Taylor, "FBW and Redundancy," p. 200.
[43] Ken Szalai, interview, Dryden Flight Research Center, 12 June 1998.

3. The pilot interface would be kept simple. It would provide access to flight-control functions rather than direct access to the computer. In fact, the DSKY wound up in a gun bay. It was accessible to the ground crew only, not the pilot.
4. Aircraft handling qualities would meet or exceed those provided with the mechanical system.[44]

Felleman was excited about the potential for fly-by-wire and was convinced Draper could meet the objectives. He and others brought into the project were riding the crest of a tsunami of self-confidence gained through their Apollo experience. Working with a small group of NASA engineers with such a focused goal was attractive. The entire program for both NASA and Draper was small, involving about 10 people in Cambridge and eventually 25 in California at any one time, with only about 10 of those concentrating on software. In contrast, at Draper the Apollo program absorbed 400 person-months' effort each month (including those temporarily assigned).[45] These two software teams quickly developed close relationships.[46]

Draper assembled a software development support infrastructure specific to the F-8. It consisted of an Apollo Guidance Computer, a DSKY, a core rope simulator, and initially an Apollo hand controller instead of stick and rudder; this mirrored the Apollo infrastructure. The hand controller caused some difficulty in testing the software, since it did not match an aircraft's controls. It was replaced by an F-8 cockpit obtained from a U.S. Marine Reserve F-8 squadron, stripped down so that the artificial horizon, altimeter, airspeed, rate of climb, thrust, g-meter, and angle-of-attack indicators were the only instruments working. Takeoff, taxi, and landings would have to wait to be done in the Iron Bird simulator at the Flight Research Center (see next section). An XDS 9300 engineering computer did transformations between parts of the simulator. Even though Draper was initially brought into the fly-by-wire program because of its expertise with the flight computer, it was overall software development capability and support infrastructure that ensured the Lab would be involved for the long haul.

Reflecting, engineers from Draper fondly remember the F-8 project because it had discrete, tangible, positive results. Vincent Megna, the program manager for Phase II, relates that working on missile guidance was significantly less rewarding.[47] Although able to see many successful test shots on such a project, Draper engineers never hoped to see actual use. On

---

[44] Dwain Deets, "Design and Development Experience with a Digital Fly-By-Wire Control System in an F-8C Airplane," in *Description and Flight Test Results of the NASA F-8 Digital Fly-By-Wire Control System* (Washington, DC: NASA TN D-7843, 9-11 July 1974), pp. 21-22.
[45] Robert R. Bairnsfather, "Man-Rated Flight Software for the F-8 DFBW Program," in *Description and Flight Test Results of the NASA F-8 Digital Fly-By-Wire Control System* (Washington, DC: NASA TN D-7843, 9-11 July 1974), p. 95.
[46] Felleman interview, 27 May 1998.
[47] Megna interview, 27 May 1998.

the other hand, developing digital fly-by-wire revolved around frequent flying and tweaking the system for increasingly better performance and reliability. The project participants sought perfection with hopes that their design would be the flight-control system of the future. Furthermore, younger engineers on the project were often taken to the Flight Research Center to present their work during reviews. This gave them exposure to the entire NASA-Draper team, a valuable experience. The Flight Research Center's participants, in turn, acquired Draper's pervasive can-do attitude and the ability to carry that confidence into later projects.

Developing the Flight Software

With the F-8 program, the Flight Research Center attempted to adopt the new role of system integrator. Previously only responsible for flight research, Jarvis' team would integrate and validate software developed by Felleman's group and ultimately decide if it was suitable for piloted flight. Draper Lab remained responsible for requirements analysis, software and interface design, simulator support, and flight-test support. Delco (AC Spark Plug when it won the Apollo computer hardware contract) also provided simulator and flight-test support, maintained flow charts of the software, and provided training on it for the Center. Hydraulics Research and Manufacturing built the secondary actuators, and LTV (formerly Ling Temco Vought) helped with the aircraft.[48]

Jarvis eventually assigned seven engineers to verify, validate, and integrate software. NASA's "Stage I" simulator for these tasks consisted of a simple breadboard with analog models for the flight-control laws. The next stage included some real hardware and better analog circuits. Finally, the team built and used the Iron Bird simulator for the work. The Iron Bird resided in a hangar at the Center for over 15 years. It was an F-8 electrically "alive," and all the hardware associated with the fly-by-wire system (computers, backup system, and actuators) was installed on it.[49] Software running in the Iron Bird would demonstrate its readiness for the actual flight hardware. In fact, the hardware in the simulator was flight-qualified and available as a spare.[50] This simulator rapidly became one of the most useful parts of the

---

[48] Philip G. Felleman, "An Aircraft Digital Fly-by-Wire System," manuscript, delivered at the 29th Annual ION Meeting, St. Louis, MO, June, 1973, p. 1; Szalai interview, Dryden Flight Research Center, 8 June 98.

[49] This was a great leap forward from previous simulators like the "iron cross" used to explore reaction controls on the X-1B. Flight hardware was not prevalent in that device, which helped pave the way for use of reaction controls on the X-15. See Edwin J. Saltzman and Theodore G. Ayers, *Selected Examples of NACA/NASA Supersonic Flight Research* (Washington, DC: NASA SP-513, 1995), pp. 16-19. Simulation advanced much further at the Flight Research Center with the X-15 and lifting-body programs, but none of the simulators for them were as sophisticated in some respects as the F-8 DFBW Iron Bird. See, e.g., R. Dale Reed with Darlene Lister, *Wingless Flight: The Lifting Body Story* (Washington, DC: NASA SP-4220, 1997), pp. 7, 27, 30, 58-59, 87, 92, 95-96, 99, 119-121, 135-136.

[50] Dwain A. Deets and K.J. Szalai, "Design and Flight Experience with a Digital Fly-By-Wire Control

program. Typically, control law development began with exercising the equations on an analog simulator, doing a linear analysis using a digital computer, coding it, and finally validating it on the Iron Bird.[51] This would help find any unforeseen shortcomings before committing to flight. For instance, engineers working with the simulator discovered that additional failure logic was needed on the various data channels. The fix was made prior to permanent rope manufacture. Both Draper Lab and the Flight Research Center increasingly relied on the Iron Bird for verification throughout the duration of the program.[52]

As the project officially got underway in January 1971, the immediate goal was to develop and verify the software that would be located in the permanent rope memory. The software had to be completed and delivered to the rope weavers at Raytheon before the year was out in order to give them enough time for manufacture before the July 1972 shutdown of the facility. The most critical time was from the March 1971 delivery of the initial specification until the mid-December goal for software release.[53] The two thousand words of erasable memory would be held for changing parameter values and any last-minute fixes. There was little expectation that the requirements definition would be other than an iterative process. No one had designed software for airborne flight control. There were several open questions that previous spacecraft experience could not answer. Cal Jarvis, Chief of the Systems Analysis Branch, Dwain Deets, and Ken Szalai (a man for whom the term "whiz kid" might have been invented) worked closely to answer those questions in the Phase I software specification that would be delivered to Draper for implementation. Szalai took the pitch axis, Deets the roll axis, and Jarvis (initially) the yaw axis, though project management duties later caused Jarvis to give up his part.[54] Ken Szalai then assumed responsibility for the yaw axis.

Deets and Szalai had worked together on a previous project. Deets came to the Center from a master's program in physics in 1962. He was assigned to a variable stability aircraft project, which was using a modified F-100 fighter. Within two years, Deets became project manager of a more ambitious experiment: converting a Lockheed JetStar corporate jet into an airborne simulator. The flight controls could be tuned such that the JetStar could act like any one of a variety of aircraft. It was an analog system, and Deets' experience with its modification increased his interest in software and made it more attractive later. Szalai trained as an electrical engineer and joined the

---

System Using Apollo Guidance System Hardware on an F-8 Aircraft," AIAA Guidance and Control Conference (AIAA Paper No. 720881, 1972), p. 4.
[51] Deets and Szalai, "Design and Flight Experience," p. 5.
[52] The Iron Bird eventually ended its illustrious career as a stationary ground target at the Navy's China Lake weapons development center.
[53] Bairnsfather, "Man-Rated Flight Software," p. 95.
[54] Cal Jarvis, telephone interview, 1 June 1998.

JetStar project fresh out of college in 1964. He actually came close to rejecting NASA's offer, initially agreeing to join the Cornell Aeronautical Laboratory. But, luckily for the Flight Research Center and NASA as a whole, he changed his mind and called, before his letter declining the offer arrived in California, to say he was coming. Szalai would later recall that the JetStar project was a real education in flight dynamics and controls. Flying in the aircraft, he could feel and see on the strip charts the effects of control system inputs and outputs. He was also able to work closely with pilots and learned how they helped shape a research program. After the project ended, Szalai had nearly a year to do some studying on other projects and coincidentally chose the Apollo guidance and navigation system.[55]

Deets and Szalai experienced difficult problems while building the software specification mainly in: 1) the use of a digital system in a previously all-analog world, and 2) the encapsulation of the computer behind an analog interface to the airplane. At the input end of the computer there was an analog-to-digital converter; at the output end, a digital-to-analog converter. When the pilot moved the stick, displacement translated to voltage. For instance, in the pitch axis, the limit of physical movement was 5.9 inches (nose up) toward the pilot and 4.35 inches (nose down) away from the pilot. The transformers were designed to generate a signal of plus or minus three volts. Therefore, the input to the analog-to-digital converter was scaled to the longer aft movement, so the forward movement had a maximum value of about 2.4 volts, while the aft movement topped out at -3.0 volts. The voltage from the transformers would be converted into bits and then serve as input to the software control laws.

At first glance, the control laws seemed straightforward. If the pilot wanted to climb, he or she added power, then pulled the stick back; the elevators then moved proportional to the stick movement. However, the process did not prove to be that simple. For instance, control devices in each axis have a deadband region in which small movements have no result. In a mechanical control system, the deadband is caused by stretching of the control cables from age and use. The deadband varies over the lifetime of the aircraft and cables. It is different in each axis and definitely unique to each airplane. Maintainers and inspectors try to minimize its effects, but in a fly-by-wire system even small discrepancies are somewhat magnified. If the fly-by-wire designers ignored the deadband, the control surface would move in accordance with every tiny motion of the stick and rudder pedals. The airplane would then be too sensitive to fly without the occurrence of pilot induced oscillations that result from constant attempts to damp motion. The deadband regions could only be determined by iteration and disciplined trial and error.

At the output end of the computer, signals causing gearing gains had to

---

[55] Szalai interview, 8 June 98.

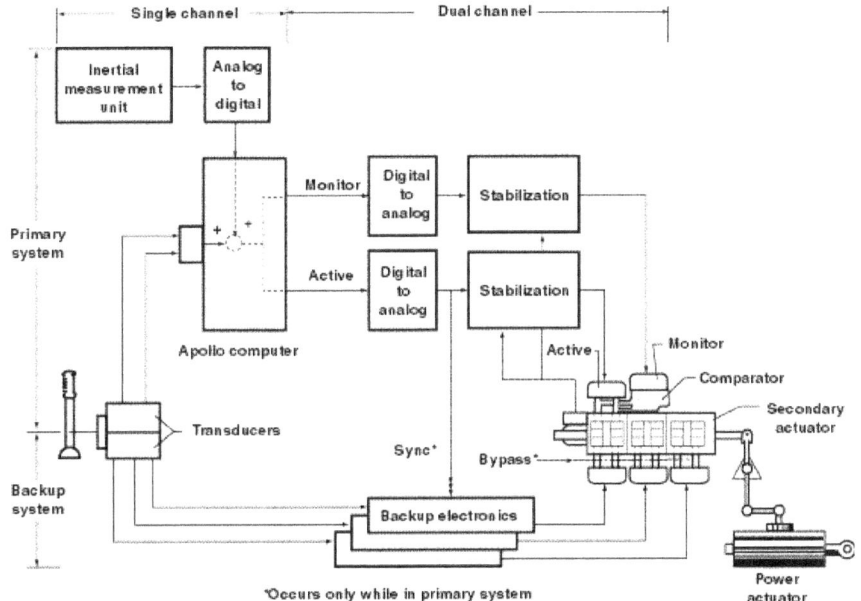

Diagram showing the F-8 Digital Fly-By-Wire Phase I system mechanization. (Taken from NASA TN D-7843, p. 33).

be properly calibrated. Movement of a control device in inches was translated by the control laws into movement of the appropriate control surface in degrees. This gearing was non-linear. In roll, small stick motions were handled sensitively, while the system translated large stick motions less sensitively. Without this accommodation, the aircraft would maneuver more violently than the structure or pilot could stand. Initially, this "stick shaping" was done by hardware, adjusting a linear variable differential transformer to provide parabolic shaping. The result was incorrect quantization of the output. This prompted the engineers to use software to shape the stick movements. The quantization problem was eventually fixed in this medium. Szalai recalled that this particular situation was another case of how you had to always keep the entire system in mind when looking at any subsystem.[56]

Equations to handle deadband and gearing were the heart of the control laws. As an example, one law in pitch was pilot trim control. The output to the actuators was a sum of the trim command from the electric trim button on the stick (often called the "coolie hat" because of its looks) and the product of stick gearing gain and the stick deflection. The stick deflection was adjusted by factoring out the deadband.

For a nose-up trim command of +2 degrees, a deadband value of 1 inch, and nominal gearing gain, a pilot trim command would come from the computer as 7.3 degrees. An analog voltage representing 7.3 degrees traveled down the two output channels (the active and a monitor) to the secondary actuators. These would cause the hydraulic actuators to move the control surface.

[56] Szalai interview, 8 June 98.

The value ranges, constants, and gains had to be determined by careful analysis and simulation. Also, since the team used a digital system, its members were required to determine a sampling rate and command quantization. Analog computers take continuous signals as input and output. Digital computers have to receive and send at discrete intervals, but fortunately, these interactions are at high speed. Every 30 milliseconds, a sampling and calculation cycle took place. Within that cycle, the mainline control calculations would be updated every 8 to 15 milliseconds.[57] The frequency of the output commands was initially too low. The first time they were tested in the Iron Bird, the stepping movements of the actuators caused a tremendous vibration. The output was smoothed by further trial and error. This is a further example of how the program could not have survived without the simulators. It is also an example of how digital systems have advantages over analog systems. The smoothing was accomplished with a pilot prefilter in the software, done before the permanent rope manufacture. This incurred no schedule delay, whereas in an analog system, hardware would have needed to be changed, a longer-term proposition.

Despite the difficult nature of these problems, the team at the Flight Research Center looked forward to applying logic encapsulated in software to the flight-control problem. As 1971 progressed, the difficulties inherent in software development became clear. Deets and Szalai reported that the use of logic "had a significantly greater impact on software complexity and verification than was anticipated."[58] Fortunately, the Draper Lab had a software architecture that simplified construction and experience with the esoteric sensors used in Phase I. No one would have purposely chosen a system for aircraft where gimbal angles have to be converted to numbers representing the speed of motion in each axis, when rate gyros provided the data directly. Gyros are devices that have a spinning element that remains oriented in space in spite of movement of an airplane or spacecraft. Gimbals, in this sense of the word, are the part of a gyro on which the spinning element is mounted. The Apollo inertial navigation system used gyros that did not have the ability to calculate rate information internally; this was left to the computer.

Exploratory work eventually led the project to the software requirements specification. This was delivered to Draper and a series of ten-hour clarifying phone calls began. The control law equations were written into the specification, arranged by axis and functional groupings, with no attempt to order them as they would be handled by the flight software. This made them implementation-independent, though more difficult to use. Although titled a "specification," this aspect made the document more of a "requirements" document than a specification. Draper Lab prepared the specification of the software from it. The variable names were cryptic and at first incomprehen-

---

[57] Felleman, "An Aircraft Digital Fly-by-Wire System," pp. 5-7.
[58] Deets and Szalai, "Design and Flight Experience," p. 2.

sible to an outsider. The following equation is an example:

$$DEC1=(KGE1)DEP1+DET1$$

DE meant "delta" or "change," C is "command," K is "constant," GE is "gearing," P is "pilot," and T is "trim." The equation can be loosely translated as: "The command change equals the gearing gain times the pilot stick position plus the change in trim." The ranges and values are located in tables at the end of the document, causing a bit of "two-finger" exercise to read it. The variable names were different for each axis, and the control laws were not "set" for a long time, though they continuously built upon each other.

Szalai says that the Draper Lab software developers "made" the F-8 program.[59] Felleman recalls Robert "Barney" Bairnsfather meticulously going over the specification, creating flow diagrams, writing code, and not even approaching a compiler until he was sure his program was right.[60] Few defects were found in his work, but getting all the numbers right was a learning experience. The software developers were limited to fixed decimal point arithmetic, which required scaling by hand to achieve the greatest accuracy, again by some trial and error.

The specification required over a dozen revisions before its final version was published in March 1973, about a month after the first version of the Phase II specification! Uncontrolled change would have destroyed both schedule and budget. The many changes to the specification were managed by a four-layer system.[61] The lowest impact were Assembly Control Board requests—relatively straightforward code changes that could be approved by the software manager at Draper. Next highest was an Anomaly—an error that needed to be repaired. Both the Center and Draper software manager signed off on it. Next was a Draper-originated Program Change Notice—during development something could not be implemented in the desired way, so the implementation had to be changed. Again, both managers signed. The highest level was a Program Change Request—a change to the specification. Both software managers and the project manager had to approve this, as there usually were schedule and budget impacts.

Bairnsfather and the others built the software using tools developed for Apollo. The assembly language system had acquired some nice features that eased long-term development, such as diagnostics, a basic and interpretive language, flexible memory allocation, cross-reference tables for variables, and the separate assembly of modules that could be integrated later. The interpretive language allowed list processing, thus making matrix arithmetic, use of vectors, and double- and even triple-precision numerical representations possible. The program was reviewed by "eyeballing effort"—a primi-

---

[59] Szalai interview, 12 June 98
[60] Felleman interview, 27 May 1998.
[61] Bairnsfather, "Man-Rated Flight Software," p. 102.

tive peer inspection—then tested via the various simulators until its release from Draper to the Flight Research Center.

Finally, a man named Al Engle got to name the software. This was a big deal since all the names of the Apollo flight software loads had something to do with the sun (SUNDISK, SUNBURST, SUNDIAL, etc.). Engle chose DIGFLY, which was supposed to be pronounced as "dig-fly." However, to his consternation, it was often mangled to "didge-fly," obliterating the intended reference to the digital system.

There were two copies of DIGFLY in the core rope.[62] It was the lone program assembly, in contrast to Apollo software with separate programs for different flight phases. DIGFLY was divided into system and application components. The system software consisted of an executive that provided task management, a restart segment that could re-initialize hardware and software in flight, and service routines to monitor the inertial measurement unit, provide self-tests, control the interface, and handle interrupts. The application software had flight control and some miscellaneous components. The flight-control portion did the mainline processing of the control laws, handled the mode and gain changes made by the pilot, and processed input from the sensors. Among the miscellaneous components were ground-test software and special-purpose applications.

Since parts of the software were similar to the Apollo code, some of it could be reused like the hardware. The display code, executive, and inertial-measurement unit alignment were taken from the Apollo 14 lunar module flight software load. The self-tests came from the Apollo pre-flight erasable memory load. Sixty percent of the eventual F-8 software was taken from Apollo.[63]

Despite the expectation that the software was correct and the hardware robust, the switchover to the analog backup flight system was carefully designed. Draper Lab used computer restarts as a solution to what were hopefully transient problems. Various logic errors could cause a restart: a parity failure in a data transfer (the bit used for parity checking was a 0 instead of 1, or vice versa), an infinite loop in the computations, an attempt to access unused memory, or silence from a running program.[64] The most famous and disconcerting restarts happened for a different cause on the first lunar landing. The computation cycle was shared by multiple programs, each getting a few milliseconds to do one cycle. The total time of the cycle exceeded 20 milliseconds, which was the limit. The computer did a restart, but the problem persisted. A nervous flight director received assurances from the computer specialists in a room adjacent to Mission Control that the restarts, though irritating, were normal. Draper wanted to allow three restarts in the F-8 before the digital channel was declared failed. As it turned out, this

---

[62] Bairnsfather, "Man-Rated Flight Software," pp. 96-97.
[63] Bairnsfather, "Man-Rated Flight Software," p. 112.
[64] Dwain Deets, "Design and Development Experience," pp. 21-22.

could take up to one second, which was extremely long for a high-speed airplane. Thus, the requirement was changed: any one failed restart would cause a switchover to the Backup Control System (BCS).[65]

There was one area of redundancy in the primary system that had to be actively managed by software. Two output channels, an active and a monitor, went to each secondary actuator. The actuators themselves were triply redundant, but that redundancy was handled by voting at the actuator itself. All of the three actuator outputs were compared one with the other, and essentially the majority ruled. In the case of the data channels, they had to be monitored for consistency. If they failed, there was an automatic switchover to the BCS.

During all of 1971, personnel, telephone calls, and paper flowed from coast to coast as the software took shape.[66] Results from testing in computer simulators and the Iron Bird were folded back into the software specification and then to the growing volume of software. In the Apollo computer, several programs could be active, each given a slot in the computing cycle. The DIGFLY was named P60. It became a source of frustration for Jarvis's software subteam, whose members were surprised by the demands of test and analysis in a flight program.[67] Line-by-line verification was the norm. As the day of rope manufacture approached, it became clearer that they had indeed gained control over the software and a full understanding of its role and capability. In the meantime, another subteam, led by James R. Phelps, was methodically converting NASA 802 to a fly-by-wire testbed for the Phase I flight program.

The Apollo hardware jammed into the F-8. The computer is partially visible in the avionics bay at the top of the fuselage behind the cockpit. Note the DSKY in the gun bay. (NASA photo E-24741).

---

[65] Deets/Szalai, "Design and Flight Experience," p. 2.
[66] Nearly 100 pages of notes from "Fly-By-Wire Telecons" are in the Dryden Flight Research Center Historical Reference Collection. Almost all the conversations were between either Dwain Deets or Ken Szalai from NASA and either Robert Bairnsfarther or Albert Engle at the Draper Laboratory.
[67] Lane Wallace interview with Cal Jarvis and Ken Szalai, Dryden Flight Research Center, 30 Aug. 1995.

Chapter Four: Converting the F-8 to Digital Fly-By-Wire

NASA's Flight Research Center was lucky enough to get a true high-performance fighter for the digital fly-by-wire program. For technology transition purposes, the military would take a demonstration of a new flight control paradigm more seriously using one of its aircraft than if the Center had used, say, a corporate jet. Commercial manufacturers perhaps would have been more easily sold by a modified Boeing 737, but at least they could consider that the flight envelope (speed, altitude, and maneuvering) on a military aircraft was larger and more difficult than anything they would immediately need. Even though the eventual aim of the digital flight-control effort was to influence the design of nearly all aircraft, it was acknowledged that the military services were the most likely early adopters.

The F-8C Crusader on the ramp at the Flight Research Center in its original livery. (NASA photo E-20095).

The Navy's specification for the Vought F-8 Crusader went out for bid nearly 20 years before NASA's fly-by-wire project started. The first carrier-based supersonic fighter, it became one of the first production aircraft of any type to exceed 1,000 miles an hour in level flight. In 1957, a then unknown Marine major named John Glenn set a speed record in an F-8, crossing the United States in three hours and 23 minutes. In the Vietnamese conflict, the F-8 became known as the "MiG Master," shooting down 19 enemy aircraft while operating off of smaller fleet carriers like the *Ticonderoga* and the *Hancock*. Some attribute its success to the four 20mm cannon it had as primary armament. While F-4 pilots were practicing long-range missile shots in their gunless interceptors/fighter-bombers, the F-8 air-superiority flyers were closing to gun range much like their brethren in World War I, World

War II, and the Korean War. Even though the sole use of guns yielded only two of the 19 kills, to many pilots the aircraft was the "last gunfighter," remembered with nostalgia by later F-14 drivers armed with 75-mile range missiles.[1] It was fortuitous for NASA that the Agency got a plane with large internal gun and ammunition bays. The new avionics boxes could easily fit in the space freed up by removal of the weapons.

The Flight Research Center received four F-8s. One was used on the Supercritical Wing experiment, one became the Iron Bird simulator, tail number 802 was converted to fly-by-wire, and number 816 was kept intact to be used by research pilots for familiarization and proficiency. The simulator aircraft arrived on 28 March 1971, and the team led by James R. Phelps immediately began opening panels and clearing the future avionics bays. One month later, on 27 April, aircraft 802 had its 37th NASA flight (the first 36 being for general pilot familiarization and not part of any specific project) and the next day went in for modification.[2] Thus, the Iron Bird conversion would stay ahead of the flying airplane conversion, providing valuable knowledge. The two aircraft sat side-by-side in the hangar, allowing technicians to move between them.

The most remarkable thing about the conversion activity was that there were no remarkable things: all the problems were manageable. As Phelps,

Members of the F-8 DFBW team moving the fuselage of the Iron Bird from under the wing after the end of the project. (Unnumbered NASA photo).

---

[1] Barrett Tillman, *MiG Master,* 2nd ed., (Annapolis, MD: Naval Institute Press, 1990); Discussions with Captain Richard Martin (ret.), who flew the F-8 in Vietnam and later commanded an F-14 squadron.
[2] James R. Phelps, personal log number 1, 28 Mar. 1971 to 8 Apr. 1972.

key avionics engineer Wilton P. Lock, and their colleagues worked diligently along, difficulties would come up and be either quickly dispatched or overcome by sheer persistence. These ranged from the usual, such as a badly fitting canopy, to the bizarre, like the sweatshirt that had flown 37 times undetected in a fuel tank on NASA missions. When participants were asked the question "What was the most difficult problem?" they uniformly answered that they could not think of one. Phelps recalled that it was like "climbing a mountain."[3] He did not feel pressure to make a certain date, which he attributes to Jarvis' management style. Everyone felt an urgency to complete work and get the airplane in the air, but not at the expense of cutting corners.

The key enabling technologies for fly-by-wire are computers, sensors, and actuators. The F-8 conversion involved all three, with two types of computers, digital and analog. In a little over a year, aircraft 802 would be internally transformed, while retaining its fighter-plane exterior looks.

Installing the Apollo Digital Computer System

The main problem with using the Apollo guidance computer and its associated systems was that it needed active cooling while it was running. It was not designed to be air-cooled, so the computer and the cooling system had to share space. The F-8 has a good-sized avionics bay behind the cockpit and above the gun bays. Removing the guns and ammunition allowed the auxiliary avionics, the DSKY, and the backup flight system to rest in the gun bays, leaving the original avionics bay for the computer and the inertial platform with the gyros—and the coolant system.

Throughout the conversion process, nothing caused a longer string of difficulties than the coolant system. The idea was to build a pallet plumbed for liquid cooling. The pallet would be shipped to Delco and the Apollo equipment installed. KECO, of Santa Ana, supplied a liquid nitrogen and ethylene glycol system that used a coolant loop to create cold sinks, which would absorb heat adjacent to the computer and inertial measurement unit. This came to be called the "glycol system" or "KECO" for short.[4] According to Phelps' log, the plumbing started in California on 11 June 1971, after Delco made final recommendations about the placement of the avionics boxes. It was shipped to Minneapolis by 25 June, and by mid-August Delco had the hardware installed and was doing vibration tests. On 16 September, the pallet arrived back at the Flight Research Center.

On the 24th, the first glitch cropped up: the cold plate inlet and outlet plumbing had been reversed, necessitating that a new line be built. The next problem was the lack of cooling endurance. Even though the optimum cooling time was shorter than desired, about an hour and a half, tests showed that the hardware stayed within temperature specifications for two hours and

---

[3] James R. Phelps, interview at Dryden Flight Research Center, 27 Mar. 1998.
[4] Phelps interview, 27 Mar. 1998.

ten minutes, longer than the expected pre-flight, flight, and post-flight total time. In January 1972, the team executed the paperwork to lower the actual specified operating time to one hour twenty minutes from one hour thirty minutes, even though the nitrogen lasted up to one hour and thirty-five minutes in some tests.[5] There are frequent notes in Phelps' logs about the coolant system troubles all the way up to the first flight. KECO sent a representative to the Center to see personally what the problems were. Ken Szalai remembers the fellow saying that the real problem was that "you guys are trying to cool the world!"[6] Ironically, after all this work, one day someone forgot to turn on the cart with power to the coolant system during a ground test. The avionics on the pallet ran for over an hour with no cooling until they triggered an overtemperature warning. No damage resulted, but once again human beings demonstrated that they are the cause of many technological failures.[7]

Pallet number one, plumbing repaired, was in the Iron Bird on 28 September 1971 and powered up two days later. The next day the flight pallet arrived, and could be installed in 802. The hardware in the Iron Bird meant that the software would be exercised in an aircraft before shipping it to Raytheon for rope manufacture. As late as July 1971, Jarvis hoped the software would be ready by late September.[8] It was fortuitous that it was not, because it could be tested yet again in the simulator with the actual hardware. The closer to the flight environment the software could get before being frozen in ferrite, the better. Even then, Jarvis' revised schedule, issued 4

The Apollo guidance computer and the inertial system on a pallet. Note tubing to carry coolant among the cold plates. (NASA photo E-23287).

---

[5] Phelps, personal log number 1, 28 Mar. 1971 to 8 Apr. 1972.
[6] Kenneth R. Szalai, interview at Dryden Flight Research Center, 12 June 1998.
[7] Phelps, personal log number 1, 28 Mar. 1971 to 8 Apr. 1972.
[8] Calvin Jarvis, personal log number 0, Jan. 1971 to June 1972.

November 1971, called for the software development to halt on the 16th, followed by a month of test, then shipment to embed it in core rope. These ropes would return toward the end of January 1972 and be used for further testing. A second version of the software, DIGFLY2, would include changes indicated by test results and some further work by Draper Lab. It was also scheduled for a month of testing before it, too, would be put into core rope, with Skylab's software the last ropes made for an Apollo computer.[9]

For pilot input to the computer system during these early days, Jarvis found two lunar module hand controllers. One had been cold-soaked in a test, so it was surplused by the Apollo Program as not usable for flight but was fine for the simulator. The second, rejected because an astronaut did not like the feel, was to be used later in 802, perhaps as a side-stick. For the ground crew and test input, they installed a DSKY in the left gun bay. When powered up, it blew out, due to a mistake in power requirements. It was replaced by a DSKY from the command module of Apollo 15, having freshly returned from the Moon.[10]

The Backup Flight System

Even though the digital system was the focus of the program, no one would fly it without backup. The analog backup system flown on the F-8 DFBW airplane was a fairly mature technology. At the Flight Research Center, analog circuitry was used as the basis for airborne simulators. Multi-threaded analog flight controls were introduced on the F-107 and RA-5C test aircraft in the 1960s.[11] The later lifting bodies test-flown by NASA had stability augmentation and control augmentation systems based on analog computers. The M2-F1 was light enough to get by with a mechanical system with no hydraulics. Designers felt that its heavier and faster successor, the M2-F2, would suffer from pilot overshoots and oscillations with these controls. Therefore they added an SAS (stability augmentation system) that sensed pilot inputs and sent opposing signals to the control surfaces. Both the pilot and the SAS had 50 percent authority in the system, so it acted as a rate damper, slowing the results of pilot movements and smoothing maneuvers. The X-24A had a triply redundant analog control system built by Sperry.[12] These projects gave the team at the Center a base of experience on which to build. The success of analog systems on the Air Force's B-47 fly-by-wire testbed aircraft and the intended use on the F-4 fly-by-wire aircraft added some confidence.[13]

Ironically, in its later production models, the F-8C already had an analog

---

[9] Phelps, personal log number 1, 28 Mar. 1971 to 8 Apr. 1972.
[10] Szalai, interview, 12 June 1998.
[11] Duane McRuer interview with Lane Wallace, Hawthorne, CA, 31 Aug. 1995.
[12] R. Dale Reed with Darlene Lister, *Wingless Flight: The Lifting Body Story* (Washington, DC: NASA SP-4220, 1997) pp. 144-145, 147.
[13] Tomayko, "Blind Faith."

system that was no longer needed with a fly-by-wire design. Called the "approach power compensator," it consisted of a computer, accelerometer, servo amplifier, and actuator, and it used the existing angle-of-attack detector. All components were in the left main wheel well except the actuator, which was in the engine bay. Its intended purpose was to maintain airspeed within plus or minus four knots on landing approaches. It could only be engaged when the aircraft was in a wing-up configuration. The pilot would set up for the approach to the carrier, and the accelerometer and angle-of-attack sensors would send signals to the computer. As long as the acceleration was one g and the angle of attack optimum, the compensator had nothing to compensate and did nothing. If the computer determined that either value changed, it had to decide whether they offset each other or intervention was necessary. If the aircraft needed a power change to maintain a good approach, the computer sent a signal to the servo amplifier that amplified it on the way to the servo actuator. The actuator moved the fuel control cross-shaft which mechanically changed engine power and throttle position.[14] It was in effect a primitive auto-throttle, a more advanced version of which is in most commercial aircraft today. The converted F-8C did not have automatic engine controls of any sort. The stability and control augmentation systems were to provide sufficient help on approach.

The system installed in the F-8C had three analog computers in the right gun bay connected to the sensors and actuators. Sperry, the analog computer supplier to several previous Center projects, produced these computers as well as the three destined for the Air Force's F-4 fly-by-wire airplane. On that aircraft, tail number 680J, they would be the primary system, and Wilt Lock thought that NASA was second in Sperry's eyes to that higher-visibility project.[15] James Morris, the Air Force counterpart of Cal Jarvis, recalled that his team ignored the NASA work.[16] Jarvis also remembers not paying much attention to the Air Force effort, since it did not use digital computers.[17] However, his logs demonstrate far more knowledge of Morris' program than his Air Force counterparts had of his. An entry for 25 August 1971 said he learned that 680J might fly as soon as mid-February, supposedly winning the fly-by-wire race for the Air Force.[18] Gary Krier recalls that the NASA and Air Force project pilots had a good relationship, with exchange of information on each other's airplanes. Jarvis and several members of his team were able to track the Air Force project quite closely. Krier remembers that there was a "race" going on, even if nobody told the people at Wright-Patterson AFB. When 680J flew on 29 April 1972, Jarvis' team was initially crestfallen. When it found out that the F-4 still had its mechanical control system, and that 680J had taken off with it, not engaging the fly-by-wire system until in level flight, they felt much better. They were convinced that their

---

[14] Naval Air Systems Command, *NATOPS Flight Manual, Navy Model F-8C and F-8K Aircraft*. NAVAIR 01-45HHC-501, 1 Jan. 1970, revised 15 Jan. 1971, p. 1-51.
[15] Wilton P. Lock interview, Dryden Flight Research Center, 25 Mar. 1998.
[16] James Morris interview, Wright-Patterson AFB, Dayton, OH, May, 1990.
[17] Calvin Jarvis interview, Palmdale, CA, 7 Jan. 1998.
[18] Jarvis notebook number 0, Jan. 1971-June 1972.

debut flight a month later would be the first time an aircraft would be completely fly-by-wire, including both primary and backup systems.[19] Ironically, Jarvis, Krier, Szalai, Lock, and Morris were unaware of the Canadian CF-105 Arrow, which flew with a dual-redundant analog system with no mechanical (or other) backup in 1959.[20]

The primary system on the F-8 had a liberal number of gates to the backup system. On any inputs that did not match up (for instance, both positive and negative trim requested simultaneously), the primary would automatically "downmode" (revert) to the backup control system (BCS). This would also happen in case of computer self-detected failures, power failure, and the loss of parallel outputs to secondary actuators (which were supposed to receive multiple copies of commands).[21] The pilot had control over the BCS engagement via three push switches on the mode panel located in the top center of the instrument panel in the aircraft. Each switch engaged one axis. The pilot could also select all three BCS axes simultaneously with a paddle switch located on the center control stick. This was primarily for emergency downmode to the BCS in the event of a significant control problem with the primary digital system. Different gains could be selected using rotary switches. To make the switchover easy, integrators in each axis tracked the primary's commands; that way they could be used as the initial orders to the system, avoiding transient spurious commands.

A card from the Backup Flight System. Note triplex voter modules at the left of the lowest row. (NASA photo ECN-7597).

---

[19] Gary Krier, telephone interview, 24 July 1998.
[20] James Morris, telephone interview, October 1995; I asked Jarvis, Krier, and Szalai about their knowledge of the CF-105 in separate telephone interviews in Feb. 1998.
[21] Philip G. Felleman, "An Aircraft Digital Fly-by-Wire System," manuscript, delivered at the 29th Annual ION Meeting, St. Louis, MO, June 1973, pp. 5-7.

At least once per flight the pilot engaged the BCS to align the inertial platform.

The fact that the BCS itself was a triplex system meant that it needed to have redundancy management. The technique used on the contemporary similar F-4 system is called "mid-value logic." The three outputs are compared and the middle voltage value is sent to the actuators. This was later used on the F-16A and F-16B, which have triple-redundant analog computers as the heart of their flight-control system.

The analog system needed a separate power supply to complete its installation, as the Apollo computer was fairly hungry in that regard. The backups to the power supply were 24-volt batteries that could keep the BCS running for an hour. These were constantly trickle-charged by the primary power supply.[22]

As with the software in the Apollo system, the computers in the BCS needed some changes that were discovered during the ground-testing phase. The problem was that changes meant hardware changes, not a few computer words. Here is where design and manufacturing efficiencies intended to help maintain Sperry's brisk computer business actually worked against a flight research program. Sperry had begun to manufacture analog computer circuits in small plug-ins containing several resistors, amplifiers, and the like, in one package. These were further installed in small boxes without much room for a technician to get to the inside. Sperry thought it was in the era of reusable parts and components. Wilt Lock points out that because of hiding individual components away, it was much more difficult to replace, say, a resistor that by itself might improve performance of the system.[23] If the circuit boards had had components that were easily accessible and replaceable, they would have been much more useful for exploring the unknown territory surrounding the F-8 conversion. Also, money could have been saved by using components from previous programs—a NASA trademark. For instance, resistors from the decade-old Dyna-Soar program found their way into the F-8. Two larger component changes also occurred: the backup systems' integrators were changed from analog to single-function digital circuits due to excessive signal drift, and filters were added to reduce noise to the actuators.[24]

While the team installed the two computer systems and tested them as diligently as the software, work went on with other parts of the aircraft. The sensors and flight controls would provide inputs to the computers; the outputs went to the actuators. All required a close connection with the original hardware.

System Inputs: Sensors and Flight Controls

It is reasonable to consider the pilot's flight controls and the sensors

---

[22] Wilton P. Lock, William R. Peterson, and Gaylon B. Whitman, "Mechanization of and Experience with a Triplex Fly-By-Wire Backup Control System," pages 41-72 in *Description and Flight Test Results of the NASA F-8 Digital Fly-By-Wire Control System* (Washington, DC: NASA TN D-7843, 1975), pp. 43, 50.
[23] Lock interview, 25 March 1998.
[24] Lock, Peterson, and Whitman, "Triplex Fly-By-Wire Backup Control System," p. 51.

together because they are both inputs to the system. In Phase I of the F-8 project, the primary sensors were those provided by the Apollo inertial measurement unit. The other inputs came from the pilot, and the original arrangement of center-stick, rudder pedals, and trim switches was kept. An addition later in the flight research program was a side-stick. Side-sticks were already used on other research aircraft such as some lifting bodies, were being installed contemporaneously in a C-141 and the 680J, and were destined for the YF-16 and the Space Shuttle.

Basically, the stick and rudder pedals mechanically positioned slider valves in the hydraulic cylinders used as actuators at each control surface to deflect the surfaces and thus maneuver the airplane. A power-control cylinder moved the control surfaces relative to the slider valve positions. This system had no feedback to the pilot, so a set of springs, bobweights, etc., was arranged to give artificial "feel" similar to that in a cable-only system. The stick would be returned to neutral once it was actuated and released. In the early days of fly-by-wire, engineers thought that such an artificial feel system would be unnecessary, probably reasoning that the electronic feedback would be sufficient for control. That was right, but pilots complained about the absence of sensory feedback from the controls. Aircraft like the CF-105 Arrow had to have a feel system installed before first flight.[25] The F-8 also had a mechanical variable gain linkage that eliminated large pitch deflections caused by small stick movements. It is interesting that this had to be recaptured in the form of "stick shaping" software when the mechanical linkage was removed. The official Navy aircraft handbook says that, "The feel forces are kept low to make the aircraft pleasant to fly and easy to maneuver."[26] F-8 pilot comments contained in the individual flight reports agree that that objective was achieved in the fly-by-wire airplane as well.

The original flight-control system also had automatic roll and yaw stabilization to improve general handling and gun platform characteristics. Loss of the stabilization system allowed the aircraft to still be controllable, but there were "drastic" reductions in maneuvering capability. The yaw damper, as with almost every jet since the B-47, and certainly on the CF-105, was the most unpleasant to lose. Immediate speed reduction and great care needed to be exercised in case of its failure. On this subject, the pilot's operating handbook for the Arrow reads almost the same as that of the F-8. The yaw axis stabilization in the Canadian aircraft is a redundant system, and in emergency mode the yaw axis is the only effective damper.[27]

To review, a pilot flying a conventional F-8 under visual flight rules would have data such as indicated airspeed, angle of attack, bank attitude,

---

[25] J. C. Floyd, "The Canadian Approach to All-Weather Interceptor Development," *Journal of the Royal Aeronautical Society*, 62, No. 576 (Dec. 1958): 845-866.
[26] Naval Air Systems Command, *NATOPS Flight Manual, Navy F-8C and F-8K.*, pp. 1-27; 4-13.
[27] *Preliminary Pilot's Operating Instructions Arrow 1* (Malton, Ontario: Avro Aircraft Limited, April 1958), pp. 22-23.

etc., displayed on the instrument panel. He or she could see that information, combine it with the view out the canopy and some physical feeling of g forces, process it in the pilot computer (a human brain), and make decisions about what to do next. These decisions would be encapsulated in stick and pedal movements that caused a change in valve positions, hydraulics, and control surfaces, which were fed back to the pilot by the feel system.

In the modified F-8, the visual cues would be present; so would the feedback; but nearly everything else changed. Once the pilot positioned the stick and rudder pedals, a completely electrical system took over. The inertial measurement unit was an arrangement of accelerometers and gyros that could track the attitude, velocity, and position changes of the airplane without depending on external devices. It served to supply the flight-control computer with a reference that was compared to the pilot's desires expressed in volts from transformers connected to stick and pedals. These linear variable differential transformers (LVDTs) were installed at the base of the stick for pitch and roll commands, but back in the tail for yaw. There were two in each axis, one for the primary, and one for the analog system. The computer (either the digital or each analog individually) would figure out control-surface position changes, and send commands on both a primary and monitor channel to the actuators, the end of the control chain.

These changes were supposed to be transparent to the pilot; they would faithfully reproduce all of the good handling qualities of the F-8, indeed make them better. There was one glitch in achieving that objective, and the NASA team found that out in a roundabout way. From 17 to 21 May 1971, Q. W. (Jerry) Burser of Delco taught a class on the digital flight control system at the Flight Research Center.[28] Ken Szalai recalled that everyone was impressed at the quality and volume of the documentation Delco produced and in having a five-day class on it. He also remembers the consternation caused by Burser's mention of "the roll rate limit." For the first time, Center engineers discovered that the design of the Apollo inertial measurement unit would limit the F-8 to a maximum of a 70-degrees-per-second roll rate as well as a 70-degree pitch attitude.[29] This was somewhat less than fighter-plane capability but would not really affect the objectives of Phase I.

The End of the Line: Actuators

The conventional F-8 had two identical power-control systems. The conversion to fly-by-wire induced a change that replaced the slider valves in these with secondary actuators. Hydraulic Research and Manufacturing of California made them for the project in 1971. They each had primary and

---

[28] Phelps, personal log number 1, 28 Mar. 1971 to 8 Apr. 1972.
[29] Szalai interview, 12 June 1998. A telephone conversation between Dwain Deets of NASA and Albert Engle of Draper Laboratory on 27 May 1971 discusses further limitations to the roll rate when gimbal angles reached a certain relationship. The notes for this are in the Dryden Historical Reference Collection.

A set of actuators is shown installed beneath the big hydraulic piston that moved the elevators on the F-8 DFBW aircraft. (NASA photo EC67-7591).

backup modes. In the primary mode the digital computer sent analog position signals for a single actuation cylinder. The cylinder was controlled by a dual, self-monitoring servovalve. One valve actually controlled the servo and the other was a model for comparison. If the position values differed by a predetermined amount, the backup was engaged. In the backup mode, three servocylinders were operated in a three-channel, force-summed arrangement. This meant that essentially the highest force value was used.[30]

Ready to Fly

The secondary actuators were the last new components installed in converting the F-8 to digital fly-by-wire. Phelps tried to upgrade the engine from a J57 P20A to the more powerful P420, but that would have to wait for later. The P20A in the aircraft was sent out to the Navy for refurbishment. It returned in October of 1971. In the fall, the avionics pallet and some other units were installed, and the software continued to be tested. In February, 1972, Phelps' team reattached the wing panels and received the core ropes with the flight version of the software. The new backup flight system came from Sperry in early March, and Delco visited to test the Apollo electronics and clear up minor problems with the inertial measurement unit. Later in the month, mechanical devices caused some difficulties: a burst hydraulic tube, an ill-fitting canopy, and fuel leaks. The engine, pilot's seat, and tail were re-installed in April.

---

[30] Robert F. Rasmussen, Tri-Tec Associates, letter to Charles L. Seacord, Honeywell Inc., "F-8 actuator evaluation of phase I," 6 June 1974.

Chapter Five: The Phase I Flight-Research Program: Digital Control Proven

All the work preparing the aircraft and software in Phase I of the digital fly-by-wire program aimed at three objectives: to gain experience with a digital system, demonstrate dissimilar redundancy (especially that synchronization is possible), and prove that a single version of the software can control an airplane reliably.[1] These objectives could be achieved early in the flight-research program. Knowing that, Jarvis was well into planning and obtaining hardware for the later phases of the project by the end of 1971, when the Phase I software was pretty much done and 802's conversion was proceeding apace. In retrospect, this made the Phase I flight-research program seem almost a rehearsal for later, more complex experiments, especially because everyone knew single-string digital hardware was unacceptable for commercial use. Nevertheless, it was a valuable series of flights in itself. Aside from easily achieving the primary objectives, the test program showed the utility of various stability augmentation schemes, incorporated a side-stick to try out the YF-16 flight-control design, and gave a half-dozen pilots fly-by-wire experience, including the chief test pilot for the YF-16. Fourteen of the 58 flight hours in 42 flights spread over 18 months from late May 1972 to late November 1973 evaluated the analog system or used it to support the YF-16. That was more than enough to convince later designers that switching back and forth to a backup was feasible.

On the first of May 1972, Bruce Peterson, project engineer and former research pilot, issued the mission rules document that would initially provide the limits and emergency procedures for the F-8 flights. The rules included the 70-

The digital fly-by-wire aircraft caught above a rare cloud layer. (NASA photo ECN-3276).

---

[1] Calvin R. Jarvis, "An Overview of NASA's Digital Fly-By-Wire Technology Development Program," in *Description and Flight Test Results of the NASA F-8 Digital Fly-By-Wire Control System* (Washington, DC: NASA TN D-7843, Feb. 1975), p. 24.

degrees-per-second roll-rate limit due to the Apollo hardware restrictions. Also, the document specified a highly conservative crosswind limitation of 10 knots, less than half that of a conventionally controlled F-8. If power were lost to the inertial measurement unit, then the pilot would fly straight and level for two minutes to let the gyros spool down so the unit could be realigned. The mission rules required most failure modes be handled by switching to the backup control system (BCS). These failure modes included: engine out, generator failure, battery voltage drop to 27 or below, and "any digital fly-by-wire abnormality"— the convenient catchall. Specific to the digital system, if the gimbal angles showed tumbling, or the computer or inertial measurement unit failure light was on for more than a minute, the devices would be turned off, automatically downmoding to the backup control system. Since the BCS was such an important safety item, its self-test had to be passed in all axes or a mission would be aborted.

The mission rules (metaphorically) in one hand, the pilot had the Cooper-Harper rating scale in the other. This is how the pilots would tell the designers how well their product worked in each maneuver. Using the scale involved three decisions about the aircraft in a particular maneuver or flight phase:

> Is it controllable?
> Is adequate performance attainable with a tolerable pilot workload?
> Is it satisfactory without improvement?

If the answer to the first question is "no," then improvement is mandatory, and the rating is 10 (the lowest). If the answer is "yes," then on to the second question, and so on. Ratings of 7 to 9 indicate major deficiencies that *require* improvement. Ratings of 4 to 6 *warrant* improvement, but could be lived with. Ratings of 3 to 1 ranged from mildly unpleasant to highly desirable performance.

Mission rules in place, pilot feedback scale adopted, hundreds of hours in the Iron Bird accomplished, Gary Krier was raring to go. Each flight, the ground crew would prepare 802 before the pilot got in the cockpit. The airplane would be connected to a power cart and a ribbon cable to a tape reader. The main software was in the core rope, but all the data needed could not be stored in the permanent memory. A flight-test program needs flexibility, so putting everything in rope would be as bad as the Sperry analog components Wilt Lock struggled with. Therefore, before every flight, the KSTART software load would be put in the erasable memory.

KSTART

The Apollo computer had only 2,000 words of erasable memory. These could be used by the software developers at the Flight Research Center to specify data constants, indicate the storage location of data to be telemetered to the ground, and even store short programs. Up to 105 variables could be adjusted for each flight. Among them, for example, were the gains in each mode and each axis. The pilot used three four-position switches on the mode control panel to select a particular gain. Position three was optimal, as well as anyone could tell before a flight. Number four was a higher gain; numbers one and two selected a lower gain. Other data in the preflight software load would be parameters to compensate for the vagaries the inertial navigation unit had in its new environment. The team kept the data in engineering terms, such as samples instead of time, and individual parameters could be changed via the DSKY. There were also branch control addresses and parameters needed for the software to jump from one place to another in the permanent memory. The downlist segment of KSTART simply listed the addresses of information needed to record and monitor the flight data on the ground, thus the major part of the telemetry stream. As flights progressed, there were eventually added three executable programs in the erasable memory: EMP-001 Restart Downmoding to BCS, EMP-004 Parabolic Stick Shaping, and EMP-007 Single Pulse Pedal Deadband.[2] These programs were in response to handling problems discovered in flight research.

The Flight Research Center developed the KSTART tape, but Felleman's group at the Draper Laboratory wrote and verified the Erasable Memory Programs. The software engineers at the Center would then integrate it. Jarvis wrote a set of strict software revision and preparation procedures for KSTART.[3] There were two diagnostic programs that helped verify the validity of the KSTART software and tape. One was called DOWNDIAG, built by Ken Szalai and Daniel Dominik, another engineer. The other was SHERLOCK, built by Szalai and his colleague Richard Maine. DOWNDIAG checked the downlink list for errors; then the list would be integrated with the constants and values and the program to form a complete tape. The entire mission load, still on punched cards at this point, would be submitted with SHERLOCK to an IBM 360 mainframe computer. SHERLOCK checked for accuracy and valid value ranges. The output from SHERLOCK listed major and minor errors. Majors were defects like invalid loading sequences, incorrect formats, unreasonable data, and number base conversion errors from octal to decimal (as required by certain hardware). Their presence prevented a flight load from being punched. Flight Research Center engineer James B. Craft wrote a program that punched the verified deck using a small Honeywell Alert computer. The output deck was duplicated and sent to the Iron Bird and MIT for checks on simulators.

Jarvis defined the mission rules for KSTART:
1. Downlink lists must have passed DOWNDIAG without error. Any changes to the list caused the deck to be resubmitted.
2. SHERLOCK must have passed without error, or with signed-off deviations.

---

[2] Robert R Bairnsfather, "Man-Rated Flight Software for the F-8 DFBW Program," in *ibid.*, p. 114.

[3] Calvin Jarvis, memo on KSTART procedures, Dryden Historical Reference Collection.

3. No flight decks could be made from changed decks that had not passed SHERLOCK.
4. Each KSTART tape was functionally verified in the Iron Bird.
5. A checksum verification called UPSUM using the DSKY was the only pre-flight erasable memory test allowed once the tape was verified.
6. UPSUM was verified by visual inspection of DSKY output.
7. A KSTART change summary was distributed to key engineers and the pilots.
8. A KSTART progress checklist was the only necessary documentation.
9. Any changes to non-Center-developed areas must have been approved by Draper Lab.

Once verified in the simulators, KSTART was recorded on Mylar tape for final loading onto the computer. An engineer would look for the correct UPSUM value on the DSKY, and then start the flight control program if it checked out. The gun bay access door would be closed, and 802 was ready for the pilot to complete pre-taxi checks.

The Pilot Checklist in the Digital Era

The especially prepared pre-flight checklist had two additional sections added due to the Apollo computer and its associated hardware. One section listed the mission rules restricting the flight envelope as a reference. The other was specific to the flight-control system. Before taxi, the servos in all three axes were tested and engaged, using switches on the cockpit control panel to the pilot's left, behind the throttle lever. The computer checkout procedure was to select direct mode, then press the computer fail switch to test the warning lights. Following that, the pilot actuated the paddle switch to go to the backup flight system. After checking these basic modes, he (all the F-8 test pilots were male) tested SAS and CAS (control aaugmentation system). The next step was to cycle the controls to full deflection, check the servos, then do the battery tests for both the primary and backup computer systems by turning off the generator. Finally, the pilot turned the generator back on and reset the servos. These procedures exercised the critical components of the flight control system that simply had to be perfect before takeoff.

Early Phase I Flights: Expanding the Envelope[4]

Bruce Peterson assigned roles for the control center crew before the first

---

[4] The following account of the Phase I flight tests is primarily based on flight reports written for each test and located at Dryden Flight Research Center. These are not complete in themselves. They are supplemented by the personal notes and interviews of Calvin Jarvis, Wilton Lock, James Phelps and Kenneth Szalai, interviews with Gary Krier, and the personal diary of Ronald J. "Joe" Wilson. From these various sources, I assembled a narrative relating to each of the 41 flights. Each flight description is derived from a collection of information from different sources.

flight. He would be the controller, the only person allowed to speak to the pilot. Project manager Cal Jarvis would keep an event log. Dwain Deets, Ken Szalai, Wilt Lock, and Jim Craft would monitor the telemetry. Team members Bill Petersen and Jim Phelps were responsible for aircraft systems. Center engineer Bruce Richardson made sure the instrumentation worked.

At 08:14:34 on 25 May 1972, Krier rotated the F-8C nose up and departed Runway 04 at Edwards Air Force Base. By 09:02:32, when his main wheels touched down on lakebed runway 18, he had achieved the chief objective of the Phase I test program: digitally controlled flight. He made two more flights in June; on all three he was limited to using the direct mode while engaged in the digital system. The flight on 8 June was general "envelope expansion" and a test of the backup system. A flight scheduled for the 16th slipped three days due to a BCS component failure that caused the preflight to be started again. This was the first supersonic flight, and there was some porpoising at 0.98 Mach. Krier did not notice any problems with quantization (delays in executing commands caused by the sampling rate) or any tendencies for pilot-induced oscillations on the first flight, but these began to crop up as the aircraft flew higher and faster. Between flights, he and the other pilots to follow would practice the mission profile in the Iron Bird for rehearsal and to look for glitches that could be repaired before flight.

The first flight attempt planned for the SAS was to be on 3 August, but frustrating glitches caused it to be pushed back until the 18th. Draper Laboratory prepared EMP-004 after the third flight to provide parabolic stick shaping, which meant that some of the handling qualities would be improved, especially in pitch. Jim Craft loaded the new KSTART tape on 2 August for the DSKY preflight. The checkout showed that new parameters for EMP-004 caused deviations in previous values for deflection of the control surfaces that were so great they required sign-off by the senior engineers. This done, 802 could fly on the 4th. When Krier started his takeoff roll, he immediately had an electrical power failure and had to abort. He cleared the runway and shut down. After towing the aircraft to the ramp, the ground crew opened the access hatch to the generator. There was lubrication oil pooled in the gear housing. To get the oil level correct, the crew removed significant amounts. Also, the cooling fan did not operate. The aircraft had its engine running for nearly an hour before taxiing out. The combination of too much oil, no cooling, and a long pre-flight killed the generator. The ground crew replaced it and the fan over the next several days. That gave the software engineers time to recheck the DSKY pre-flight on 9 August. This time the results were worse. An "increasingly noticeable aileron oscillation" in both ailerons caused the engineers to terminate the pre-flight.

By the 18th, the project team had done enough fixes to proceed with the flight. When Krier taxied out to the runway, he called Bruce Peterson on the radio saying, "I have some good news and some bad news. The good news first: the digital system and the electrical system are OK." Peterson asked, "What's the

The mode control panel mounted on the top center of the front instrument panel where the gunsight used to be was the most frequently used interface with the flight computers. Pilots could select different modes in each axis with the push buttons and manage the gains with the rotary switches. The panel also had warning lights to indicate computer, power, or inertial measurement unit failures. (NASA photo E-24742).

bad news?" Krier replied, "I have a flat tire." Normally, this would not be much of a problem, but the flying schedule was tight that day, with the F-8 flight sandwiched between two F-111 tests. The ground team had to do a tire change as fast as an Indy 500 pit crew to get sufficient flight time. Pilots for a Convair CV-990 down from NASA's Ames Research Center who were following the day's tests on the radio heard the exchange between Krier and Peterson. Sensing a chance to sneak into the schedule, they fired up their engines and sent their telemetry team to storm the ground control center. The defending F-8 telemetry team refused to give up their places over a cut tire. Krier called to the 990 pilots as he taxied by, "You could at least wait until the corpse is dead!"

The SAS flight did go successfully that day. Krier reported pitch handling much improved, and in roll, setting up stable bank angles was much easier. Only four days later he was in the cockpit again for a slightly more ambitious test of the SAS and new stick gearing gains. Szalai issued a flight readiness report dated 22 August 1972 that defined the differences from the previous version of KSTART so everyone would know the situation with the fixes. Takeoff was with the SAS engaged at the low-gain setting of 1 in each axis. Krier reported a smoother ride at takeoff than in direct mode. His chase pilot in an F-104 reported turbulence on climb-out (they were flying in mid-afternoon, when the heat of the day boils the air at low altitudes). The F-8 handled the turbulence with aplomb. Level at 20,000 feet and flying at 300 knots, Krier tried SAS in all axes at different gain settings. Stability increased as the gains increased, with yaw and pitch damping best at a setting of 3. After some turns, he tried the pitch CAS in all gains. Attempting formation flight back in SAS mode, Krier gave the F-8 a Cooper-Harper rating of 5 due to considerable attention needed in the roll axis. A pilot induced oscillation (PIO) started in roll and had to be stopped. He tried to increase speed to 400 knots, but it proved to be too fast for the system as it was configured. All axes became too sensitive, so Krier eased off to 350 knots. After a few more tests, he set up for a landing on lakebed runway 33 instead of the concrete runway due to wind gusts from the north. He set 802 down safely despite sudden jumps and drops in the airspeed, and with the ailerons and rudder full-over to overcome the crosswind. In his report, Krier said the touchdown took place in considerable side drift and under "marginal control." Privately, he said he was essentially out of control when the wheels reached the runway.

Between this flight with its harrowing ending and the next, there was a special Open House at the Flight Research Center in (somewhat early) celebration of the 25th anniversary of the XS-1's breaking the sound barrier on 14 October 1947. The F-8 was on display for the entire weekend of 8-9 September. After Phelps' team towed the airplane back to the hangar, they discovered "Curt" scratched on the right rear fuselage! Jim Phelps reported that the airplane crew left the graffiti as a reminder to be very vigilant when a one-of-a-kind research aircraft is on public display.[5]

More important repairs were in progress on the formation flying problems. On the 15th, Krier tried out partnering again with an F-104. The roll axis remained troublesome. There was a noticeable lag between stick movement and aircraft movement. A preliminary analysis attached to the flight report said that there was a delay of 280 milliseconds. The delay time attributed to the digital system was 105 milliseconds. Trying to figure out what else was contributing to the problem was hampered by the fact that the measurement instrumentation had a sampling rate larger than the digital flight-control system's quantization, so intermediate data was being lost.

NASA flight 44 of 802 was the first by a pilot other than Krier. Thomas McMurtry, the chief pilot on the Supercritical Wing project, which used another modified F-8, flew on 21 September 1972. He was accustomed to using the

---

[5] James Phelps, review notes on the manuscript of this book, Oct. 1998.

nose-wheel steering to keep the airplane centered on the runway on takeoff, switching to rudder control at about 60 knots. He tried to do the same with 802, but when he switched to the rudder he found it much less responsive than on the Supercritical Wing aircraft. Distracted by having to make large rudder inputs, McMurtry did not achieve a smooth rotation. When the airplane left the ground, he found himself in a lateral PIO. He stopped it, but it returned and had to be stopped again. As a result he rated both takeoff and climbout worse than the basic F-8. After exercising the control system at altitude, McMurtry returned for a landing with no wind. However, there was some convective heating close to the ground, causing the wings to suddenly dip or rise. Compensating too rapidly for this took McMurtry close to another lateral PIO. As he approached the runway, he tried very hard to keep the wings level, setting him up for another PIO. Why did McMurtry have different PIO problems than Krier? As accomplished aeronautical engineer Duane McRuer has pointed out, there has never been a fly-by-wire system that flew without pilots experiencing PIOs in early tests.[6] Also, McMurtry had much less time in the Iron Bird. His thorough familiarization course included six hours of class time devoted to the hardware, control laws, normal and abnormal operation, and 27 hours in the simulator, compared to Krier's involvement in the program from the beginning and over 200 hours in the simulator. The engineers still found out that the handling characteristics had to be improved for pilots to step into a fly-by-wire aircraft as comfortably as they did moving among conventionally controlled ones.

The next six flights were all evaluations of handling in SAS, CAS, and BCS modes. Gary Krier flew one in October and one in early November, then two in December. Tom McMurtry piloted one flight in early December and the first flight of 1973 on the 10th of January. Problems in both October and November reduced the flight frequency. In October, the team grounded 802 due to the backup system's failing its self-test and also some troubles with one axis. After the early November flight, there was a fuel leak.

On 30 January, Krier flew a mission in the F-8 Digital Fly-By-Wire aircraft to compare its performance with that of the F-8 Supercritical Wing. Once the two aircraft were in formation, engineers measured fuel flow, exhaust gas temperature, and engine revolutions per minute (RPMs) on the fly-by-wire aircraft while at Mach 0.9 and four intermediate steps up to Mach 1.005. Krier then performed a few ground controlled approaches (GCAs) down to 200 feet when he returned to Edwards. These are more commonly available at military airfields than civilian, and they require an experienced radar operator. An airplane with no precision approach equipment on board can achieve a precision approach by having the GCA controller call maneuvers based on radar-derived height and distance information. These are rarely flown to altitudes lower than 200 feet—if a pilot can not see the runway at that height, it is best to break off the landing and fly elsewhere. Typical commands are "three degrees left," and "100 feet low."

---

[6] Duane McRuer, interview with Lane Wallace, Hawthorne, CA, 31 Aug. 1995.

GCAs are good for checking out an airplane's ability to remain stable at low speeds with frequent small maneuvers. The F-8 DFBW handled them well.

The only flight in February aborted due to an error in loading the KSTART tape that caused an infinite loop—the software was stuck repeating the same few instructions over and over and could not progress. It turned out that one item was not entered before starting the load. This flight began a series of tests of what happens when control surfaces moved more violently than normal, such as in dogfight maneuvers, and also of tracking stability using a gunsight Bruce Peterson got from a Marine A-4.[7] The three flights in March all had these objectives.

On 6 April 1973, McMurtry and Krier flew a two-flight sequence that revealed anomalies in the digital flight-control system. It took nearly three weeks for engineers to complete the analysis that explained what happened and why. The first problem was an aileron offset. After landing and reaching the NASA

The F-8 DFBW leads the F-8 Supercritical Wing in formation flying exercises in late January 1973. (NASA photo ECN-3495).

ramp, McMurtry started the shutdown procedures that would end flight 56. He used a modified checklist that left the primary computer running for the following flight. When he switched off the telemetry master switch, the ailerons moved to an asymmetrical position that would cause a roll in flight. The crew chief asked McMurtry to cycle the stick, and the control system continuously gener-

---

[7] It had to be installed lying on its side due to the mode control panel's occupying the original location and the general lack of real estate on the main panel. James Phelps, review notes on the manuscript of this book, Oct. 1998.

ated a hard left roll signal. Ken Szalai and his software team went to work. First Szalai found out exactly what McMurtry had done in the cockpit. Then the team tried to replicate the failure in the airplane in a controlled environment on the ground with all the data recorders running. It took two days, but they were finally successful on 10 April. Even then, it only happened once during several on/off cycles of the telemetry master switch. When the ailerons offset, Szalai's team dumped the erasable memory. With the printout and the telemetry data, they spent a week doing analysis. On 16 April, a taxi test confirmed their suspicions. The telemetry master switch had little to do with the problem. The answer lay in software.

There was a Roll Rate Command (RRC) mode in the system that limited roll rates. On flight 56, an integrator was disabled in the RRC logic. This simplified the RRC mode to be a roll rate damper for that flight. When the integrator was in place, the software checked for output outside pre-set limits and subtracted the excess from the integrator output. The integrator bias remained even after the stick was moved to neutral. Since the integrator was disabled, the output was equal to the integrator bias, and the aileron offset was the result. The pilot moved the stick right or left, moving one aileron up and the other down. When he returned the stick to neutral, the ailerons remained where they were instead of returning to neutral. Thus, the limiting logic should not have been used with the integrator off. A change to the KSTART tape setting a constant to zero effectively eliminated the limiting logic. Still, the team did not know the cause but felt it was safe to fly.

Apparently the engineers believed this without doing the testing, since Gary Krier made flight 57 three hours after the aileron offset showed up on 56. The ground crew noted a roll rate drift in the backup mode prior to takeoff but allowed Krier to fly anyway. Everything was all right until three seconds before touchdown, when the RRC mode suddenly switched to direct. Krier had to make a lot more stick inputs in direct mode, exacerbated by 35-knot winds. These rapid inputs reduced the hydraulic pressure in both power-control channels so much the entire system downmoded to the BCS as the wheels touched the runway. James Craft was assigned to analyze this downmoding problem. He was able to replicate the low hydraulic pressure and suggested a simple fix of either increasing the hydraulic pump capacity or keeping engine RPMs higher on approach.

Krier flew all three flights in May. These primarily tested stability and control using roll steps and rudder pulses with some pitch evaluation in different gain settings. There were more low-speed handling tests using ground-controlled approaches. Aircraft 802's 60th flight as a NASA plane—23rd of the digital fly-by-wire program—ended the first year of tests in late May of 1973.

Back in early March, Phelps had heard that there would be only seven to nine more flights in the program. At that time, Jarvis planned to have an all-digital Phase II with three or four computers in the primary system. As an intermediate step, there would be a Phase IB using only two primary computers.

This was still not enough for commercial use, as a failure would initially cause confusion over which machine had failed, but solving that problem would be interesting and a step forward. Jarvis projected the first flight of Phase IB in December 1974. Felleman and some associates came out from Draper Laboratory on 28 February and 1 March 1973 for Phase II planning sessions. They projected software development for Phase IB would have to begin immediately to make the 1974 date. Over the next month the initial software specification for Phase II arrived at MIT. Phase IB then more or less disappeared because Jarvis decided to go directly to Phase II. (This decision is discussed more completely in Chapter Six.)

Early in April, he authorized the purchase of a side-stick for $4,000 and the training of four additional pilots in July. The YF-16 program decided to use an analog fly-by-wire system and wanted to use a side-stick. Gary Krier and Bruce Peterson flew to Forth Worth at General Dynamics' invitation to try out the new flight-control system for the YF-16 in a simulator. They gave their professional evaluation at the plant. As they got into their T-38 for the flight back to Edwards, the F-16 project pilot escorting them asked for an informal evaluation. They told him that the system had real problems. General Dynamics asked the NASA program to try out the side-stick to see if it was a good decision, and to let some of the future YF-16 pilots fly with it.

In the meantime, there were seven flights of the F-8 DFBW in June and July to meticulously exercise different combinations of gearing gains at both low and high speeds. Continued fixes for minor discrepancies held up the use of other pilots. This was fortunate, because it was better to have an experienced pilot like McMurtry on board when a power-control hydraulic tube burst, causing an emergency landing. The tube was supposed to be .065 inches thick, but somehow there was a .035-inch tube. In the meantime, the side-stick arrived and 802 did not fly for two months while it was installed.

Flights with the Side-stick

There were experiments with side-sticks in other fly-by-wire programs.[8] The major advantages of a side-stick were eliminating an obstruction to seeing the instrument panel and making it easier to design a reclined seat for better g-tolerance. There is never an abundance of real estate in a fighter cockpit, where instruments and switches are jammed together and mounted at least down to the pilot's knees. As the amount of avionics increased, this could only get worse. Hence, the invention of multifunction display screens that collected data together. These, with small print and large amounts of information compressed on screens, must be seen more clearly than analog instruments. Getting the stick out of the way would allow easy views of all the forward-mounted instruments and place both the pilot's hands near to the side-mounted switches. Also, there would

---

[8] See, e.g., R. Dale Reed with Darlene Lister, *Wingless Flight: The Lifting Body Story* (Washington, DC: NASA SP-4220, 1997), p. 152.

be symmetry in the cockpit. The pilot would have the control stick in his right hand, throttle in his left, both festooned with switches and buttons. In combat, the pilot would sit reclined, both hands able to toggle all needed functions, looking out the window with a head-up display showing critical air data.

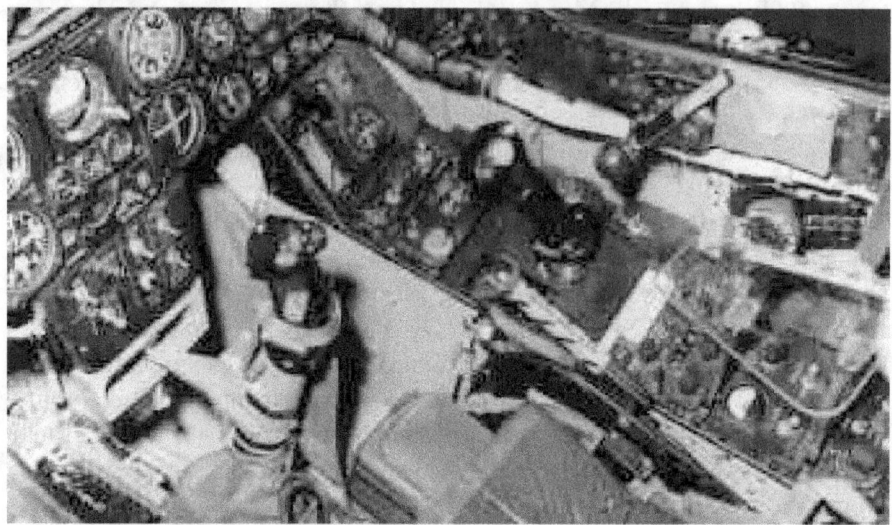

The side-stick was mounted on the right instrument panel. Note the clear plexiglass armrest that could be folded up for stowage. The pilot could still read the instruments under it by lifting his arm a little and looking through the transparent material. (NASA photo E-26466).

A side-stick could be installed much more simply with a fly-by-wire control system than with a mechanical system. There were two schools of thought on which type to install. The first type was a force-sensing side-stick. The pilot, wanting to climb, pulled back on the stick, like a conventional one, but it did not move. Instead, force sensors would translate the pressure applied by the pilot into a voltage. This signal would then be transmitted to the computer. This type of side-stick often caused the "Popeye" effect on the pilots. Like the cartoon character Popeye the Sailor Man, they experienced their forearm growing larger as they subconsciously did isometrics with the stick in violent maneuvering. The other type was a displacement stick. This side-stick did move, but not much. The pilot would get some feel, but the distances it could be moved would not be more than one-eighth to one-quarter inch in any direction. The side-stick chosen for the YF-16 was of the displacement type.

The F-8 installation put the side-stick on the right side of the cockpit, and connected it to the analog BCS only, meaning that Wilt Lock would have primary responsibility to make it work. Jim Phelps worked with Gary Krier on stick placement and an arm rest. Krier requested that the grip be mounted nearly vertically, instead of at the forward angle used in other installations. The arm rest could be folded away so the range of motion of the pilot's arm would not be inhibited when using the center stick, as well as to uncover some switches on the right side instrument panel. On 13 August 1973 the side-stick installation began, and it ended on the 28th. By 19 September, Krier and the aircraft were ready.

Peterson added to the mission rules document for the first side-stick flight. He had goals of safety and fidelity to the YF-16. Takeoff and landing would be with the center stick, as the floor for side-stick operations was 5,000 feet. There was a 4-g limit on maneuvers. The control in both pitch and roll in the YF-16 was proportional; hence, it was like an undamped F-8. Consequently, the CAS or RRC mode could not be used.

Taxiing to the runway the day of the flight, Krier briefly engaged the BCS and enabled the side-stick. He tested the effect of taxi feedback and made some relatively sharp S-turns. The system felt normal, so he switched back to the primary system and the center stick. After takeoff and climb to 20,000 feet, he changed to backup mode and enabled the side-stick at 250 knots. There was the usual momentary transient when the BCS came on line, but no transient due to the side-stick. Krier rated his initial pitch and roll maneuvers with the new control as a 2 on the Cooper-Harper scale. He noted that the adjustment to the side-stick was quite rapid and that it behaved better in the roll axis than indicated by work in the Iron Bird.[9] Increasing air speed through 275 to 300 knots, Krier found that the stick became more sensitive at higher speeds. He tried out various common maneuvers for the next hour: instrument flying, approaches (to an offset of the ground at 15,000 feet), turns, missed approaches, and so forth. All rated 2s except the pitch and heading control was a 1 and only holding a 2-g turn rated a 3. Krier felt that roll acceleration using the side-stick was actually better than using the center stick in the BCS mode. He reported no forearm fatigue after the one hour flight.

Items added to flight rules for the next flight on 25 September 1973 included one that required the yaw SAS mode to be selected for any side-stick landings or takeoffs and another that set the wind limits at 10 knots down the runway and at 5 knots of crosswind. That opened the envelope to those critical maneuvers. Toward the end of the flight, Krier did a couple low approaches over the runway and was satisfied enough to land with the side-stick. On 3 October the flight plan called for a side-stick take off and also a landing if the wind was calm. Krier took off successfully with the side-stick enabled and did some GCA approaches. The same day Tom McMurtry got his first chance to fly with the side-stick controller. Penciled in at the end of the typed mission rules for that flight was a restriction that if there were a servo that could not be reset or if its circuit breaker tripped more than once, then the aircraft was to return and land in digital mode. Obviously, something related to the servos had come up. The procedure was gone from the mission rules for the next flight, so a fix evidently was made. McMurtry got to do a side-stick takeoff on this flight, and the goals continued to be exercising the BCS/side-stick combination. He also flew the final side-stick evaluation flight on 17 October. In six flights, three by each pilot, NASA proved

---

[9] The author, a private pilot who had never flown an aircraft with a stick, tried out a side-stick in an F-16 simulator with high fidelity in the control system. It was surprising how easy it was to adjust from a control yoke to the side-stick. It was also very comfortable to use, once I figured out the rate limits and did not try to be Popeye.

the YF-16 control scheme workable, and it was incorporated into the new aircraft.

Pilot Familiarizations

From 24 October to 27 November 1973, four new pilots flew the final six flights of Phase I of the F-8 digital fly-by-wire program. By that time, almost all anomalies or inconsistencies were purged from the system, so a pilot relatively unfamiliar with it could use it successfully on the first flight. On the 24 October mission, Phillip Oestricher, General Dynamics' chief test pilot of the YF-16, took two rides in NASA 802. He had nothing but praise for the flight-control system and the NASA team, whom he called a "bunch of real pros." On his first flight, he used the digital system for takeoff and for most of the time in the air. Oestricher's ratings for nearly all maneuvers were 1 or 2s. He felt the airplane handled better than the conventional F-8. He used the side-stick toward the end of the morning flight, and landed with it. After a turnaround of only two and one-quarter hours (Oestricher said his Reserve unit would take four hours), he was rolling on a side-stick takeoff. Oestricher devoted the entire mission to side-stick checkout. During the debriefing, he emphasized the comfort of the side-stick in the F-8, with its slight cant. The YF-16's stick had to be more vertical because there would not have been enough room for the pilot's thumb if it had been canted.

A few months later, Oestricher was the pilot of the unscheduled first flight of the YF-16. The flight test program for that aircraft was to begin in February 1974. On 20 January, he was doing a high speed taxi test at Edwards when suddenly the airplane rolled and the wing dragged on the ground. Instead of reducing power and running off the runway, possibly damaging the aircraft, he added power and overcame some sickening pitch and roll oscillations. After keeping close to the runway on a go-around, he landed without incident. Those who saw this performance praised his quick reflexes and outstanding flying ability. However, perhaps his previous takeoffs and flights in a fly-by-wire airplane with a side-stick helped a little as well.

On 31 October and 8 November 1973 the former X-15 pilot, William H. Dana, got his chance to try out the F-8. Einar Enevoldson flew on 19 November, and the Phase I program ended with astronaut Kenneth Mattingly's familiarization flight on 27 November. There were many more requests for rides, especially from the YF-17 program pilots, but a project decision ceased familiarization flights in favor of beginning the transition to Phase II.[10]

On to Phase II

When NASA 802 returned to the hangar after the Phase I flights, there still

---

[10] Kenneth Szalai, personal notes, 16 Nov. 1973.

was an important area of concern regarding fly-by-wire: since it was an electrical system, what effect would a big spike like a lightning bolt have on it? General Electric won a contract to find out. The chief worry was that a lightning strike would cause electro-magnetic effects on all channels of a fly-by-wire system at once, thus eliminating the positive effects of redundancy. GE used a transient analysis technique developed at NASA's Lewis Research Center. An electric transient of the same waveform as, but a much lower amplitude than, a typical lightning strike was pumped into the body of the aircraft. Then the results were extrapolated up to lightning level. These strike tests showed that there would be some damage but that the system would never fail because of them.[11] Thus assured, Jarvis' team continued Phase II planning.

Two events during the research program helped the digital fly-by-wire team gain some visibility. On 16 November 1972, it received a NASA group achievement award. Then the first week in March 1973, Gary Krier appeared before the members of the House Committee on Science and Astronautics to tell them that fly-by-wire had proved viable and was worth future support to make it robust enough to attract the attention of commercial users. That was one goal of Phase II of the program, but, as with the YF-16, this valuable national asset would also be used to help out another high profile program: the Space Shuttle.

The Lightning Technologies equipment sparks during a lightning test of the F-8 DFBW. (NASA photo E-35496).

---

[11] J. Anderson Plumer, Wilbert A. Malloy, and James B. Craft, "The Effects of Lightning on Digital Flight Control Systems," in *Description and Flight Test Results of the NASA F-8 Digital Fly-By-Wire Control System*, pp. 74-75, 80-81.

Chapter Six: Phase Shifting: Digital Redundancy and Space Shuttle Support

Phase II of the F-8 Digital Fly-By-Wire program lasted for roughly 12 years beginning in 1973. The basic objective of Phase I was verifying the feasibility of flight control using a digital computer. For Phase II, it was flight control using multiple, lighter weight, production-quality digital computers with adequate collective safety and reliability, i.e. a more practical digital flight-control configuration. Soon at least two other major objectives began to receive attention. The Langley Research Center proposed using the aircraft to test adaptive and analytic control laws (control laws that self-modify depending on conditions). So one objective for the multi-computer version of the F-8 became solving a key issue for future fly-by-wire aircraft: would control laws or redundancy management be the chief difficulty in design?[1] Another objective was support of the Space Shuttle flight-control system's development. A key NASA scientist involved in the design of that system was Dr. Kenneth Cox of the Manned Spacecraft Center (renamed the Lyndon B. Johnson Space Center in 1973). He had alertly followed what was happening at the Flight Research Center and kept Calvin Jarvis' team informed of, and as advisors to, decisions made about the Shuttle. The two programs eventually adopted the same processor, enabling the Shuttle to reap the rewards of a hardware shakedown whose value far exceeded the money it contributed to the F-8 project. Later, the F-8 program would help solve an embarrassing and dangerous pilot induced oscillation problem for the Shuttle program. After achieving the objectives of Phase II, the F-8 program moved into a Phase IIB which consisted of a series of experiments in sensor analytic redundancy, resident backup software, further work on remotely augmented vehicle studies by Langley, and other experiments.

The Space Shuttle gained immediate benefits from NASA's fly-by-wire research. Here Endeavour returns from space to the Dryden Flight Research Center. (NASA photo EC92-05165-2 by Carla Thomas).

---

[1] Kenneth J. Szalai interview, Dryden Flight Research Center, 8 June 1998.

As noted above, the first version of the Phase II software specification was issued in early March 1973, before the final version of the Phase I specification and almost nine months before the last Phase I flight.[2] Project manager Jarvis' notes are filled with Phase II procurement activities from 1972 on. At the beginning, plans were to fly a multi-computer system by late 1974. The actual first flight of Phase II was in late August 1976, two years and nine months after the last flight of Phase I. Why did it take so long? Largely, because it was almost a totally new system. There were new computer mounts, the removal of the KECO cooling system, new computers (both primary and backup), a new three-channel interface unit, a modified mode and gain panel, a computer interface unit for the pilot, bigger pumps for the hydraulics, removal of the inertial system and its replacement by rate gyros and linear accelerometers (more typical of aircraft flight control instrumentation), inclusion of the side-stick as part of the primary system, more powerful actuators, a new generator, new fuel tanks, and an upgraded engine. This was only the hardware. There also was new software with pioneering redundancy management functions and software for ground control of the airplane in the Remotely Augmented Vehicle experiments, which meant even more hardware, such as a command receiver and dedicated downlink. Nearly everything inside the skin of the aircraft was changed, and the only external evidence of the modifications consisted of a new antenna and a tail-mounted video camera. These changes took time, and the new hardware was not always working perfectly when installed, necessitating rework. Ken Szalai recalls that the software development took "longer than anticipated," a common problem even with today's new programs.[3] Furthermore, it took nearly a year to verify the original release and its updates. (The first flight of Phase II used Release 7A, Mod 1, indicating many revisions).[4] Szalai remembers going outside for a break during yet another all-night debugging session and looking at the starry sky of the high desert to try to clear his head. There was a lot of overtime preparing for the Phase II flights. At first, Jarvis wanted to make a less ambitious transition to avoid some of the negative side effects of the "big bang" approach in which all the changes would come before first flight.

The Short-Lived Phase IB

Jarvis announced at a meeting on 16 February 1973 that there would be a Phase IB in which the current aircraft would have two primary computers installed. This would enable trying out computer synchronization on an

---

[2] Dwain A. Deets and Kenneth J. Szalai, "Phase I F8DFBW Software Specification," Final Revision, on, 30 Mar. 1973, and Kenneth J. Szalai and Joseph Gera, "F-8 Digital Fly-By-Wire Phase II Software Specification," Revision F, 1 Dec. 1975.
[3] Szalai interview, 8 June 1998.
[4] Ronald J. "Joe" Wilson, interview, Dryden Flight Research Center, 10 June 1998.

easier configuration than the three or more machines needed in Phase II. The first flight of this system was set for December of 1974, about a year after the last flight of Phase I and a year and nine months after the software specification arrived at the Charles Stark Draper Laboratory. He also said he would be going to Langley to discuss Phase II before the end of the month.[5] The two-computer system with the backup was quite safe, with complete system failure probabilities very low, on the order of one chance in a billion.

Soon this version of Phase IB disappeared. Langley engineers opposed having more than one computer on the aircraft in the next phase since they favored the single-channel, high-reliability approach demonstrated by the Apollo space program. This was obviously unacceptable to the Flight Research Center, as it obviated the goal of testing multi-computer synchronization and redundancy management, which was felt to be a necessity for cost-effective aircraft flight-control applications. Pressure from NASA Headquarters to settle the issues in addition to funding constraints had caused Jarvis' team to propose the two-computer system as a compromise.[6] Later, when the Space Shuttle program promised funding to support flight-controls development, Phase IB quietly went away, and a three-computer system became the basis for Phase II.

Phase IB returned briefly in 1975 when General Electric contractors doing a new lightning study referred to a IB that consisted of a single one of the new computers installed in the Iron Bird.[7] Except for that reference, the numerology returned to a general designation of Phase II. Peter Kurzhals from NASA Headquarters expressed an opinion that all two-computer experiments be dropped. Jarvis and Szalai concurred.[8] Phase II later split into A and B sub-phases in which IIA represented the general proof of the redundant multi-computer concept and the Shuttle support flights. Phase IIB was a series of discrete experiments to extend the proven system.

Finding an Airplane

New computers would be an obvious part of the next stage of the program. Jarvis also wanted a new aircraft that would be more suited to the objective of demonstrating a multi-computer flight-control system to commercial aircraft manufacturers. Characteristics such as twin engines or even the idea of using a small transport like the DC-9 were attractive. But the first airplane considered was a "target of opportunity." In June 1972, Jarvis heard that the Air Force's F-4 fly-by-wire project was cancelled. He wanted to meet with project manager James Morris and discuss incorporating it in the NASA

---

[5] James Phelps, logbook number 3, 16 Feb. 1973.
[6] Calvin Jarvis, telephone interview, 19 Sept. 1998.
[7] F.A. Fisher, "Lightning Considerations on the NASA F-8, Phase II Digital Fly-By-Wire System," (Pittsfield, MA: General Electric, June 1975).
[8] Kenneth Szalai, personal notes, 14 Mar. 1974 and 25 Mar. 1974.

program as an option to Phase 1B.⁹ The aircraft was at Edwards AFB for testing, and Gary Krier got to fly it from the back seat in July (the Air Force version of the F-4 had controls in both cockpits, the Navy version only in the front cockpit).¹⁰ There was a restriction that no fly-by-wire takeoffs and landings would be done during the NASA courtesy flight, a capability retained with the original mechanical control system. On 19 July 1972 Jarvis met with F-4 (Project 680J) personnel to discuss aspects of a possible joint program.¹¹ However, the Air Force soon received more funding for the project, and the F-4 was made statically unstable by the addition of canards. It kept its analog-computer-based flight-control system and flew for several more years.

The F-4 that the Air Force converted to an analog fly-by-wire system was considered for Phase II of NASA's project. More funding for the Air Force appeared, however, and the canards were added to make the airplane longitudinally unstable. (U.S. Air Force photo).

In late July 1972, Raymond Hood of Langley added to the list of candidate aircraft a Learjet that Boeing would modify, a 737, T-39, and DC-9. Later in the summer the RB-66 and a new aircraft, the S-3A, were considered, with Langley sending an engineer to check out the S-3.¹² All these aircraft were twin-engined. However, a transport category aircraft would require an FAA Supplementary Type Certificate. Documentation and testing to receive one would cost about $1,000,000. Just in case, James Phelps was asked to check on the availability of spares for the F-8. He found out that the reconnaissance F-8s would fly until 1980, the fleet and reserve H and J models until 1983.¹³ This was long enough for the planned closure of Phase II in late 1978. The project proceeded with the idea that the F-8 would fly early in Phase II, and a converted Lockheed JetStar NASA already owned

---

⁹ Calvin Jarvis notebook number 0, Jan. 1971-June 1972.
¹⁰ Gary Krier, telephone interview, 24 July 1998.
¹¹ Calvin Jarvis, notebook number1, 19 July 1972.
¹² Jarvis, notebook number 1, 20 July, 22 Aug., and 22 Sept. 1972.
¹³ Phelps, logbook number 3, 20 Feb. 1973.

would be modified to replace it later. The cost of the JetStar's modifications (including adding pilot safety devices like ejection seats) was estimated at three million dollars. In contrast, the F-8 modifications would be $850,000.[14] NASA ultimately decided to stay with the known aircraft and the lower cost for the entire program.[15]

Finding a Computer

Choosing a computer for Phase II took nearly two years. The Flight Research Center, Draper Laboratory, and Langley worked closely on the decision. In December 1971, Philip Felleman of Draper visited with computer specifications. A NOVA computer with four thousand words of memory would cost $20,000. The Honeywell 601 with the same memory was attractive at less than 30 pounds. The RCA 215 and Control Data Corporation Alpha were also on the list. The Alpha's price was $35,000 each for a lot of 25 computers.[16] All of these machines were relatively inexpensive and lightweight, but they were short of memory and overall processing power. The next group of more powerful computers included the Honeywell 801 "Alert," the Sperry 1819A, and Langley's nomination, the General Electric CP-32A.[17] These machines typically had larger memories of about sixteen thousand words, longer computer word sizes such as 18 bits, and a steeper price tag in the $70,000 range per computer.

In the meantime, the Shuttle program was trying to choose a machine as well. Its engineers had no experience base with aircraft control systems, as both the Gemini and Apollo vehicles were fly-by-wire but not designed to be very maneuverable in the atmosphere. So, they had to get information from aeronautical projects. As early as October 1971, Howard W. "Bill" Tindall wanted to bring the Shuttle flight-control designers out to California to see what they could learn from the F-8 project. Tindall was the primary NASA liaison to Draper during the Apollo program. On 4 May 1972, John "Jack" Garman, a key NASA engineer on the Apollo flight controls and later a manager of Shuttle software development, called Ken Szalai. Szalai briefed Garman on failure analysis, software control, flight software readiness, and mission rules for the F-8. Early in 1972, William McMahon of the Manned Spacecraft Center told Jarvis that the Shuttle program thought about buying the Sperry backup system.[18] It was still searching for a suitable computer for the primary system.

Draper Laboratory not only suggested computers, but also attributes that its engineers thought an aerospace machine should have. The software

---

[14] James Phelps, logbook number 4, 25 Jan. 1974.
[15] Jarvis interview, 19 Sept. 1998.
[16] Jarvis, notebook number 0, Jan. 1971-June 1972.
[17] Jarvis, notebook number 0, 6 Jan. 1972.
[18] Jarvis, notebook number 0, 18 Feb. 1972.

engineers there wanted a machine that used floating-point arithmetic. The Apollo computer was a fixed-decimal-point machine. It represented real numbers with the decimal point always in the same location, so to have more value to the left or right of the point meant hand-scaling the number—a difficult and error-prone process. Floating-point machines stored the value of the number and the location of its decimal point. Arithmetic is easier to program in floating-point, but it requires a larger computer word size and is not as accurate as fixed-point when the latter is done well. However, large word sizes meant more accuracy in general.

Draper also wanted a high-level programming language for the software. This required a compiler that would take the source code of the program and translate it into the machine's unique low-level code. These low-level "assembly" codes execute faster but are much more difficult to write and maintain than those in higher level languages. The high-level languages like FORTRAN had been around for 15 years, and Felleman's group wanted to take advantage of them. Richard Parten, NASA's Shuttle software development manager, authorized an experiment in which two teams of equal ability coded a function in both assembly language and a high-level language on the same computer. The high-level language code was slower and less compact, as expected, but in all aspects no more than 15 per cent less efficient than the assembly code. Parten authorized using high-level languages for the Shuttle, reasoning that the performance margin was too slim to make it worth the extra maintenance costs. The Manned Spacecraft Center had already considered a language called HAL. It was an extension of FORTRAN that made it possible to do vector arithmetic and schedule real-time programs, among other good features. Draper wanted to use it as well. The software engineers would have to wait until the computer choice could be made to find out if a compiler for HAL would be available on that machine.[19]

In the meantime, Jarvis zeroed in on some computers already used in aircraft to run avionics other than flight controls. The Autonetics D-216, a 16-bit-word, 16-thousand-word-memory machine was in the Rockwell B-1A and cost about $85,000.[20] It quickly went out of consideration as machines with larger memories entered the contest. More memory would be needed to store the additional synchronization and redundancy management code for a triplex system, especially if a high-level language were chosen.

On 5 October 1972, Flight Research Center and Langley Research Center engineers held a then-Phase IB coordination meeting at Langley. The team in Virginia found a new computer: the Singer-Kearfott SKC-2000. It had floating-point arithmetic, and the memory was 16 thousand words,

---

[19] The news about Draper Laboratory's desires is in a memo between Albert Engel of the Laboratory and Ken Szalai of the Flight Research Center, 10 Aug. 1972 (Dryden Flight Research Center History Office). The story of the choice of HAL and a description of the language are in James E. Tomayko, *Computers in Spaceflight* (Washington, DC: NASA CR-182505, March 1998), pp. 109-110 and Appendix I II.
[20] Jarvis, notebook number 0, 23 June 1972.

expandable to 24 thousand words.[21] However, it would cost $185,000 in the larger configuration. The Flight Research Center wanted the Teledyne 43M, which would only be about $50,000. Since the Center wanted to buy three machines each for the Iron Bird and the F-8, plus a development machine or two for Draper and maybe a spare, the cost difference was roughly a half million dollars for Teledyne computers versus nearly two million for Singer-Kearfotts. Also, the SKC-2000s would be on an extended delivery schedule, while Teledyne offered to loan a machine to Draper right away. The groups at the meeting decided to select the class of computer by 1 November, and the actual machine by 1 December.[22]

After the Phase IB coordination meeting on 6 November 1972, Jarvis told Ken Cox in Houston and Dr. John Bird at Langley that Draper would set up meetings with both the Singer-Kearfott and Teledyne representatives to set final pricing and specifications.[23] Almost simultaneously, the Shuttle program found the IBM AP-101 computer and it had its first mention in Jarvis' notes on the 10th of November.[24] Ken Szalai's notes reveal that the IBM computer had a 32-thousand-word memory, consumed 370 watts of power, and weighed 47.7 pounds. In contrast, the TDY-43M was in two boxes of 35 pounds each, and the SKC-2000 drew 430 watts, and weighed 90 pounds.[25] By 28 November, Jim Phelps was studying the interface needed between the F-8 and AP-101, and whether it would be suitable for a quad-computer installation as well as for the proposed Control Configured Vehicle experiment that would add a canard to the aircraft. No external cooling would be needed for any of the three computers, but the IBM internal blowers were good to 50,000 feet while above 30,000 feet the Singer machine needed ram air from an intake that did not yet exist. Also, the IBM and Teledyne computers could use the current power generator, but the SKC-2000 would need a new, larger one. Finally, the SKC machine was so big Phelps dreaded the stuffing job necessary to fit multiple computers into the space allocated. If a canard were installed, its structural needs in the computer bay would make it impossible to use Singer-Kearfott's computer.[26]

On the last day of November 1972, IBM sent Vincent Obsitnik, Gib Vandling, and Edward Zola to the Flight Research Center to present the AP-101.[27] Even though all three computers were thoroughly discussed by year's end, the 1 December deadline for selection passed and nothing definite happened for several months. By the first of March 1973, at roughly the time

---

[21] Buying an additional eight thousand words of memory for that machine was $15,000, or about a $1.88 per word. In the summer of 1998, I purchased 32 megabytes of additional memory for the machine I am typing on at this moment. It cost $48, or about *one ten-thousandth of a cent* per word.
[22] Calvin Jarvis memo to multiple addressees, Dryden History Office, 5 Oct. 1972; Kenneth Szalai, personal notes, Dryden History Office.
[23] Calvin Jarvis memo to multiple addressees, Dryden History Office, 6 Nov. 1972.
[24] Jarvis, notebook number 1, 10 Nov. 1972.
[25] Kenneth Szalai, personal notes, Dryden History Office, 11 Nov. 1972.
[26] James Phelps memo to Calvin Jarvis, 28 Nov. 1972.
[27] Jarvis, notebook number 1, 30 Nov. 1972.

The AP-101s like those intended for fly-by-wire need an interface unit. Here is one largely handmade for the Iron Bird. Note the triple redundancy. (NASA photo ECN-5067).

the initial version of the software specification went to Draper, the Flight Research Center set 12 March as the issue date for a formal request for proposals . Five weeks later the bids were due, with evaluation taking a couple months. The winner had to deliver the initial lot of five computers by September, four more by March 1974, one for the Iron Bird by 1 April 1974, and the first one to Langley by June 1974.[28]

During the final competition, the major players in the Shuttle flight-

---

[28] Szalai, personal notes, 1 Mar. 1973.

control system effort campaigned for the AP-101. Jarvis got a call from William Zimmerman of Intermetrics (developers of the HAL compiler) saying that if the Center chose the same computer as the Shuttle, the compiler and support software it was developing would essentially be free. Garman's boss in Houston, James Satterfield, said that any software built for the Shuttle would be made available. John Miller of Intermetrics said that his company was suggesting changes to the AP-101 to make executing HAL more efficient and that these changes would be in the computers delivered to the F-8 project since they were already being developed for the Shuttle.[29]

If the goal of Shuttle support were to be met, and, more importantly for the short term, if the Shuttle program would send $1 million to the Flight Research Center to augment joint program funding, the choice was clear. The engineers at the Manned Spacecraft Center essentially settled on the AP-101, even though it did not take actual delivery for a while. On 27 August 1973, IBM signed the contract to supply the computers to the fly-by-wire program.[30] Vincent Megna, who led large portions of the Phase II software development at Draper, said that the final decision was NASA's, even though the Laboratory was involved at every step.[31] HAL was abandoned due to incompatibilities with the final Shuttle hardware arrangement and the difficulty of software conversion, but IBM did make significant changes to the AP-101, forced by poor design and construction.[32] By taking the lead in the use of the computers, the Flight Research Center repaid the Shuttle program several times over by uncovering discrepancies that would have hurt the Space Transportation System's development schedule.

AP-101 Woes

The overall hardware arrangement for the Phase II system was simpler conceptually, but more difficult to implement, than the Apollo system used in Phase I. The new pallet mounted three computers and a single three-channel interface unit. The computers delivered to the program could do both floating-point and fixed-point arithmetic. They could process 480,000 instructions per second, about sixty times faster than the IBM computer used on the Gemini spacecraft less than a decade earlier. The initial prediction was that the machines would have a 5,494-hour mean time between failures.[33] It turned out that the actual figure was much lower, causing Phil Felleman to remark later that fixing the AP-101 ranked as one of the major results of the program, equal to proving the concept of redundant systems.[34]

---

[29] Jarvis, notebook number 1, 10 Apr. 1973; 21 Aug. 1973; 12 Sept. 1973.
[30] Phelps, log number 3, 27 Aug. 1973.
[31] Vincent Megna, interview, Draper Laboratory, 27 May 1998.
[32] Calvin Jarvis, notebook number 2, 12 Sept. 1974; Jarvis, interview, Lancaster, CA, 7 Jan. 1998.
[33] IBM, AP-101 Technical Description, 7 Oct. 1974, Owego, NY.
[34] Phillip Felleman, interview, Draper Laboratory, 27 May 1998.

The disappointments started soon after deliveries in 1974. The F-8 program eventually received serial numbers one through nine of the AP-101, which, of course, were the most likely to fail. The bugs at first were fairly localized: all the floating-point features failed on number one, an instruction did not work on number two, and this was coupled with a 50 percent drop in the data transfer rate.[35] By the middle of 1975, failures became endemic. There were 19 faults in the first seven computers, yielding a 204-hour mean time between failures, less than five percent of the projection. At times only one computer was working. There was no pattern to the bugs apparent at that time.[36]

Finally, the engineers found a common problem. The circuit boards, built in layers, had separated during use and caused no end of short circuits and other failures. It turned out that the manufacturers of the boards used a watered down coating fluid so it would be easier to spread on. It was so thin it seeped between layers and expanded when heated by the running computer.[37] IBM rectified this and other manufacturing errors, but often the fixes were applied to a particular machine and not all. By late 1975, IBM delivered all computers, but the modifications were different, depending on the reliability. Here is Jarvis' table from 16 October, where the higher the modification level, the more extensive the fixes:[38]

Serial Number	Modification Level
1	4
2	9
3	7
4	7
5	5
6	5
7	5
8	7
9	9

An indication of the effect on the program is that the first computer cost $87,000 and the last $130,000.[39] As the first flight approached in late summer of 1976, heat-related problems such as cracked ceramics and seals continued, so IBM sent John Christensen and Fred Hudson out to California to baby-sit the machines and provide faster service. The frequent failures slowed software development and verification.

Ken Szalai described what was projected to happen and what actually

---

[35] Jarvis, notebook number 2, 1 Aug. 1974.
[36] Jarvis, notebook number 2, 7 May 1975.
[37] Megna interview, 27 May 98; Szalai interview, 8 June 1998.
[38] Jarvis, notebook number 3, 16 Oct. 1975.
[39] James Phelps, log number 6, *passim*.

happened with the AP-101s during an in-house seminar on redundancy management in 1979.[40] The mean-time-between-failures record of the AP-101s was expected to rise to 1,000 hours at 10,000 hours of cumulative operation and stay level for a while. On delivery of the flight system, things were pretty much as expected: 675 hours between failures at 7,000 hours of use. But from February to July 1976, when operations reached 1,200 hours per month, the mean time between failures declined to 500 hours. Following final installation in the aircraft at 13,500 cumulative hours, the failure interval hit a low point of 375 hours in September 1976. It was only after major rework, which delayed the flight research program by four months, that it eventually recovered to 500 hours by 18,000 cumulative hours. By mid-1978, the mean time between failures reached 750 hours after 25,000 hours of operation. At no time did the computers meet, let alone exceed, IBM's reliability projections.

Software

While the F-8 underwent extensive hardware modifications, the Draper Laboratory and Ken Szalai's software team at the Center solved the problem of how to mechanize a multicomputer system to act like a single computer for control laws and like three independent computers for fault tolerance. The Phase II software specification showed how much the F-8 program team

During the long hiatus between the Phase I and Phase II flights of the F-8 DFBW, Boeing responded to the Air Force's call for a short-field takeoff-and-landing cargo plane with the triple-redundant digitally controlled YC-14. Thus, it was the first multi-computer digital fly-by-wire aircraft in the air. (Photo courtesy of Boeing).

---

[40] Kenneth Szalai, "Redundancy Management," slides from in-house seminar series, 29 June 1979.

learned from Phase I. It was a different document in tone and content. There was a more narrative style and it contained suggestions about software development itself.[41] The software would be scheduled to execute cyclically—a Draper Lab trademark—within a twenty-millisecond loop. This made the synchronization easier to accomplish. The computers would attempt to synchronize up to three times, sending each other discrete signals and sharing data, the attempts totaling no more than 200 microseconds. If a computer failed to keep in step, the others would ignore it and move on. A failed computer's self-test software would probably detect the failure and restart it. In contrast, IBM programmed the primary Shuttle system of four computers as an asynchronous, priority interrupt system.[42] This meant that the highest priority module of the software loaded in a particular mission phase would run until complete, then the next-highest priority module would run, and so on. To synchronize, each computer halted every time one of three actions occurred: an input, an output, or a change of module. The computers would exchange messages telling each other what they had just done. Miscomparisons or a failure to exchange information in four microseconds indicated a failure. This is a more complex method for synchronization.

The software for Phase II was also larger than that for Phase I due to a need to handle new pilot interface devices. The mode-and-gain panel still had the direct (DIR), stability-augmentation (SAS), and control-augmentation modes (CAS), but added a maneuver-load-control button in CAS mode, which was the predictive augmentation in pitch, and expanded the four-position gain switches to five. There was also a digital autopilot with Mach-hold, altitude-hold, and heading-hold selections. Finally, the Phase II system allowed pilot access to the computer software from inside the cockpit through a Computer Interface Panel. This contained three seven-segment displays, two thumb wheels with numbers zero to nine, and "enter" and "clear" buttons. The pilots mostly used it to start pre-flight self-tests and to initiate the Remotely Augmented Vehicle mode, in which the aircraft's control laws would be resident on a ground-based computer and commands for the actuators sent up.

Internally, the software used fixed-point arithmetic for sensor data and floating-point for the control laws. The memory layout of the software started with two thousand words of data; then the operating system and redundancy management comprised the next three thousand words. There were five thousand words for the control laws; sensor redundancy management took up about 2,500 words; preflight tests, about the same amount of space; and the ground display and program load instructions occupied the final four thousand words of space. This totaled 19,000 words of the initial buy of 24,000-word memories expandable to 32,000 words.

---

[41] Kenneth Szalai and Joseph Gera, "F-8 Digital Fly-By-Wire Phase II Software Specification," Revision F, 1 Dec. 1975.

[42] James E. Tomayko, *Computers in Spaceflight*, Chapter Four.

Draper Lab designed the 20-millisecond inner loop to begin with the master interrupt; then the synchronization, computer-redundancy management, scheduler, and input read took place in the first two milliseconds. The sensor signal selection occupied the next three milliseconds. The first control law execution took two milliseconds, followed by the outputs to the actuators and the displays. The second control law and signal selection came next, then data recording and the Computer-Interface-Panel read. The remaining six or so milliseconds were devoted to self-tests. The signal-selection-sensor redundancy management worked this way: if there were three good signals, the system would take the mid-value; with two good, it would take the average; one good became the default.[43]

Early in 1976, the software matured to the point where the pilots could be in the loop for verification. Draper Lab intended Release 5 of the software to be the flight release and scheduled a flight qualification review on 22 June 1976. However, it arrived stillborn at what was now officially the Hugh L. Dryden Flight Research Center. There were frequent unexplained restarts. The qualification meeting was cancelled, and in the last week of June over 2,500 power-up starts tested changes in the procedures designed to eliminate the problem. Release 6 arrived early in July configured as a flight tape. A pilot used it right away and reported no anomalies. Draper Lab issued Release 7 by the end of July with the flight qualification meeting rescheduled for 10 August.

The flight qualification meeting reviewed the test philosophy and responsibilities. Six engineers were on Szalai's verification team: Richard Larson, Sam R. Brown, Ronald "Joe" Wilson, Kevin Petersen, Richard Glover, and Szalai. They went over the tests they conducted, the results coming in graphical form and in a series of one-page Software Verification Reports that chronicled the test objectives, set-up arrangements, results, and conclusions for each test. A year of seemingly endless work had finally paid off. The direct and stability-augmentation modes were ready for flight, but the control-augmentation mode and the autopilot needed more testing.[44] These unfinished components were not needed on the first flight.

The Computer Bypass System and New Actuators

A lot of overtime went into obtaining and installing the revised analog backup system and upgraded secondary actuators. These were the domain of Wilton P. Lock. He wanted more powerful actuators for Phase II. The backup had a name change to the Computer Bypass System, and he also wanted better computers to match.

---

[43] Kenneth Szalai, Phillip Felleman, and Joseph Gera, "Design and Test Experience with a Triply Redundant Digital Fly-by-Wire Control System," AIAA paper 76-1911, delivered at the AIAA Guidance and Control Conference, San Diego, CA, 16-18 Aug. 1976.

[44] Memo, "F-8 Digital Fly-By-Wire Software Qualification and Review Meeting," 10 Aug. 1976.

It was thought that failures due to the new primary system might cause increased use of the backup, reason enough to obtain newer technology. Sperry offered a more up-to-date version of the original equipment for $355,000, or it could supply the same technology as used on the Air Force fly-by-wire program using the 680J F-4 aircraft. This system had only three avionics boxes instead of six, and would only cost another $45,000.[45] Lock chose it.

The modifications of the actuators increased the power of the system. One change made the signals from the analog computers to be "force summed" when they reached the actuators, resulting in a quicker response.[46] The redesigned secondary actuators produced 20 percent more force due to an enlarged piston. The new actuators were also more reliable: Hydraulic Research modified the Phase I hydraulic actuators to provide a purer triple redundancy. In Phase I, there were only two hydraulic sources for the actuators. For Phase II, there were three, neatly matching up with the three primary and secondary computers. The secondary electric actuators had three channels, one dedicated to each computer in the primary system.[47] The actuators were also shared by the analog computer bypass system in the event of a total failure of the primary digital system—one that never subsequently occurred in operation. These two subsystems completed the list of components needed for the new fly-by-wire package.

The triply-redundant Computer Bypass System mounted in a gun bay of the F-8C. (NASA photo ECN-5223).

---

[45] James Phelps, log number 5, 20 Jan. 1975.
[46] Sperry Flight Systems, "Description and Theory of Operation of the Computer Bypass System for the NASA F-8 Digital Fly-By-Wire Control System," Phoenix, Dec. 1976.
[47] Hydraulic Research, "Final Design Review Phase II Servoactuators," May 1975.

Preparing the F-8 for Flight

While the ensemble of fly-by-wire hardware and software was assembled, Jim Phelps's airplane-conversion team prepared the F-8 for installation and did some modifications of its own. Jarvis held a system preliminary design review on 18 and 19 September 1973 that resulted in final approval of the cockpit panels, so they could be fabricated. At that time, the team expected the software and Sperry's analog system modifications to be the pacing items, finished by late September 1974. The lightning tests, sneak-circuit analysis (see below), computer deliveries, Iron Bird modifications, and failure analysis would all be complete by mid-1974. Final qualification and first flight would be by late December 1974.[48] As we have seen, computer hardware and software delays set the program back at least a year and a half, allowing a less hectic pace than the Phase I modifications on the aircraft itself. Nevertheless, Phelps was already looking at connector and gyro placement for Phase II while the Phase I flights were still going on.

The new pallet and general hardware preliminary design review took place on 28 and 29 November 1973, right after the last flight of Phase I. The triplex computer system and the new interface unit would pretty much fill the avionics bay behind the pilot. More of the electronics boxes would wind up in the gun bays. On 23 January 1974, George Quinn of Draper Lab brought out mockups of a pallet and an attitude-gyro mount. There were no major discrepancies, so Phelps authorized fabrication.[49]

By 1975, the F-8 area of the hangar again looked much like it did in early 1972, during the first cycle of modifications. Crews moved between the simulator and aircraft 802 doing their work. According to Phelps's notes at that time, there were only eight NASA mechanics and technicians and ten engineers on the project, less than 20 core persons of a team that would top 50 when receiving a Group Achievement Award in late 1977.[50] Gary Krier could get into the refurbished Iron Bird and exercise the preliminary software

---

[48] Memo from Calvin Jarvis to multiple addressees, 24 Sept. 1973.
[49] Phelps, log number 4, 23 Jan. 74.
[50] Phelps, log number 5, 6 May 1975:

F-8 Team:	Mechanics/Technicians	Engineers
	James D. Hankins	James Phelps
	Willard E. Dives	James Craft
	Gene Webber	Kevin Petersen
	Walter P. Redman	Michael R. Earls, part time
	Harvey B. Price	Calvin Jarvis
	Carl R. Ajirogi	Kenneth Szalai
	Alfred R. White	Ronald J. "Joe" Wilson
	William J. Bastow	R. Bruce Richardson
		Gary Krier
		Wilton P. Lock

The Dryden Flight Research Center *X-Press* of 16 Dec. 1977 listed over 50 names of NASA employees receiving the Group Achievement Award for fly-by-wire.

releases. The new 680J-type analog system boxes arrived and were installed. Another nagging problem cropped up: F-8s on the USS *Hancock* had nose gear strut failures after about 200 landings—attributed to a deck modification that changed the landing geometry when the aircraft caught the number one arresting wire on the carrier with its tail hook. Would it happen on concrete?[51] It never did, but it was worrisome.

The final design review for the entire Phase II system took place on 28 and 29 May 1975. Forty-eight people were present, including representatives from Johnson Space Center and Rockwell for the Shuttle program. The agenda shows software reviews filling the entire first day. The second day, Sperry had an hour on the computer bypass system; Hydraulic Research a half-hour on actuator redesign; Francis A. Fisher of General Electric one hour on lightning tests; Rockwell an hour for a Shuttle status review; the Flight Research Center an hour to review the aircraft modification status and instrumentation; the meeting ending with Langley's hour on advanced control laws. Aside from reports of a few more hardware and software delays caused by changes to the system design and continued AP-101 problems, the program appeared in good shape.[52] Draper Lab reported that the Phase II software development supported the Shuttle Backup Flight System work. It too had a cyclic operating system.[53]

Fisher's lightning study and recommendations had less impact than one would expect with an electronic control system. He found that magnetic fields leaked into the interior of the aircraft, and he made some suggestions for sealing the leaks. However, leaks were inevitable on an airplane with so many openings in the fuselage. The Phase II configuration had the same lightning characteristics as Phase I, so Fisher recommended simply to avoid flying into or near thunderstorms. This is good advice for pilots flying airplanes with mechanical systems as well. There were no lightning tests on the AP-101 computers themselves, just static discharge tests. Fisher said it was not worth the cost to make the processors lightning-resistant.[54]

Another interesting study done in support of Phase II was "sneak circuit analysis." Boeing's Houston office did this complex work from April 1975 to March 1976.[55] A "sneak" is a combination of conditions that cause an unplanned event without a hardware failure. These types of failures are rarely caught in system testing and do not have a clear cause-effect relationship. For instance, in an electrical network such as the one on the F-8, a "sneak path" had data or energy going along an unexpected route. "Sneak timing" would be energy- or data-flow or a function inhibited at an unsuspected time. These are the most difficult defects to fix after a system is fielded; therefore, doing

---

[51] Phelps, log number 5, 12 Feb., 7 Apr. 1975.
[52] Calvin Jarvis, memo to multiple addressees, "Final Design Review for Phase II System," 4 June 1975.
[53] James E. Tomayko, *Computers in Spaceflight*, Chapter Four.
[54] F. A. Fisher, "Lightning Considerations on the NASA F-8."
[55] Boeing, "Sneak Circuit Analysis of F-8 Digital Fly-by-Wire Aircraft," D2-118582-1, Mar. 1976.

the work parallel with design is much more effective. Boeing found 76 probable sneak-circuit instances, recommended 12 design changes, and found 468 mostly insignificant drawing errors in the documents it examined.

As 1975 wound down, the Iron Bird pallet had the flight configuration of the hardware installed. Wilt Lock was working nights for months with Sperry engineers trying to get the computer bypass system operational. Center Director David Scott announced in a "State of the Center" talk on 15 December that the entire F-8 program would end in 1978.[56] That turned out to be a premature statement.

As the year turned, Jarvis set the first flight of Phase II for 10 May 1976 and the final push by all hands began. Ken Szalai's weekly reports reveal both urgency and thoroughness as he chronicled thousands of tests along with frustrating glitches and heroic overwork. In January the team lost two weeks because of a failure of the Release 2 software to synchronize the computers.[57] The flight pallet delivery date was 27 February, then 22 March, and it finally arrived on 29 March. The team quickly took advantage of the arrival: Gary Krier flew the simulator with the flight pallet on 5 April and it earned ratings of between 1.5 and 3 on the Cooper scale.[58] Even so, the May flight date was clearly gone, and 10 August became the new target.

During the third week of June, Krier flew the profile for the first flight in the Iron Bird, noting several actuator anomalies and transients. One part of the team fixed these while another part concentrated on the software problems with Releases 6 and 7. Three days after the flight qualification review on 20 August, Krier made the first high-speed taxi tests. Then on 27 August, he departed Edwards AFB in the F-8 for the first flight of Phase II.[59]

Looking back on the development of the Phase II hardware and software, the participants agree that they underestimated the system's redundancy-management effort but that it was "workable." They also complain about inadequate tools for system development. The word "agony" frequently appears in their remarks. Most conclude, though, that there were few other problems and that they had a sound design philosophy. The wide variety of uses of the Phase II system—Shuttle support, flight in a remotely augmented vehicle, experiments in sensor redundancy management, resident backup software, etc.—supports their belief. Krier's first flight—however delayed— was a tribute to their hard work, perseverance in spite of adversity, and teamwork.

---

[56] Phelps, log number 6, 15 Dec. 1975.
[57] Kenneth Szalai, Weekly Reports, 9, 16, and 30 Jan. 1976.
[58] James Phelps, log number 6, 28 Jan., 8 Mar., and 5 Apr. 1976.
[59] Calvin Jarvis, notebook number 4, 21 June, 27 Aug. 1976.

Chapter Seven: The Phase II Flight-Research Program: Proof of Concept, Space Shuttle Support, and Advanced Experiments

The Phase II flight-test program began on 27 August 1976 and ended when Edward Schneider landed flight 169 on 16 December 1985. In between, using the converted F-8, the Dryden Flight Research Center proved that multiple digital computers could be used for flight control, found out the optimum data sampling rate for the Space Shuttle, demonstrated that an airplane could be flown with control laws operating on the ground, and demonstrated advanced redundancy management techniques for sensors and other important technologies developed from computer-controlled flight. There were some close calls due to computer failures and a runaway pilot-induced oscillation, but the 169 flights all ended without any damage to the aircraft. The project team designed the flight-control system to be robust and to survive problems without a hiccup. There is an impression lingering barely below the surface that they *wanted* a few computer glitches to see how well the redundancy management worked. There was always the proven Computer Bypass System (CBS) as a backup.

The converted F-8 and the Space Shuttle prototype Enterprise share the ramp in an early assemblage of fly-by-wire aerospacecraft. (NASA photo EC78-9354).

The Phase I mission rules, which guided pilot actions in case of failures, had to be changed for Phase II. There had to be some idea of what to do in case of single-channel failures now that there were three channels. Going directly to the Bypass System all the time was not acceptable. Eventually, this kind of flight-control system had to be installed in commercial aircraft with no backup. The rules basically came down to a two-part procedure: try a reset of the indicator; then, if a problem persists, return to base in a configuration governed by this table:

Failure in:	If in:	Return in:
Primary	Primary	Primary
Primary	CBS	CBS
CBS	Primary	Primary
CBS	CBS	Primary

Rules seemed hardly necessary for chief project pilot Gary Krier. He always had a plan, actually several, in case of any possible emergency. Joe Wilson recorded one instance in his diary. On 19 August 1976, there was a technical briefing for the first flight with some people present who were outside the F-8 project team. Milton Thompson and Bruce Peterson, two highly accomplished pilots, questioned Krier about his intended abort runways for the test. He quickly answered that he would use runway 04 for departure with a limit of a 15-knot tailwind and a 5-knot crosswind. He would not use 22 because sufficient lakebed runways did not exist off the departure end for an emergency landing. If there were an immediate abort at low altitude, he would turn right to land on runway 17. If the winds were so high from the tail that 17 would have too great a crosswind, he would come left to 27. If he had enough altitude and speed at the abort, he would make the approach to runway 18, a little farther away.[1] Of course, all this would be in front of his mind as he rolled down the runway for takeoff, and he had to decide to abandon a particular emergency runway as winds, speed, and altitude changed.

Getting started the day of the first flight was a problem. The Computer Bypass System failed its self-test twice in a row, so the preflight was restarted. During it, all the monitor lights in the control room on top of Building 4800 illuminated, even though the corresponding panel lights in the aircraft were lit normally. A canopy latch also malfunctioned. Krier and ground crewman James Hankins determined that the latch problem did not affect ejection or pressurization, so they made a manual fix and pressed on.[2] Shortly thereafter, everything righted, Krier made a *déjà vu* takeoff from runway 04 in the primary direct mode, did each planned maneuver, checked out each axis of the CBS, and 45 minutes later landed on lakebed runway 18. James B. Craft made sure the three "best" of the nine computers were in the

---

[1] Ronald J. "Joe" Wilson, Flight Notes Diary, 1975-1977, 19 Aug. 1976, Dryden Historical Reference Collection.
[2] Wilson Diary, 1975-1977, 27 Aug. 1976.

aircraft for the flight: Serial number three with 2,135 hours of operation was Channel A; number eight, 1,576 hours, Channel B; and number four, 2,951 hours, Channel C.[3]

For the second flight on 17 September, computer number four moved to Channel B and number seven replaced it in Channel C. This flight was a watershed for the program. All three computers failed at some time during the test procedures, one of them in flight. The objective of this mission was "envelope expansion," increasing the range of altitude, speed, and g forces the F-8's new control system could withstand. The intended maneuvers included sustained flight over 40,000 feet to see how the cooling system would work in a low-air-density environment, sustained supersonic flight at 20,000 feet to check how the cooling system handled moderate heating, and 4 g turns or higher to see if any mechanical problems would crop up.

Gary Krier used the afterburner for takeoff due to the full load of fuel needed to achieve the high altitude and speed required for the research flight. After a normal climb to 20,000 feet, Krier did some small maneuvers to exercise all the control surfaces. Then he repeated those maneuvers at plus-50-knot-speed intervals, eventually using the afterburner again to nudge the F-8 past 500 knots, supersonic speed. He did stability and control tests at up to 527 knots, Mach 1.1, then began a supersonic climbing turn to 40,000 feet in afterburner. Twenty-three minutes after takeoff, trying to level off, Krier cut the afterburner at Mach 1.21, and within one second the Channel A fail light and its associated air-data light illuminated. The computer tried a restart, failed, then just quit. Without hesitation, Krier began to return to base, following the rules and staying in primary mode since he was in it when the failure occurred. An uneventful landing on the two good computers followed.[4]

The Dryden Flight Research Center team was collectively reaching the end of its patience with the AP-101s. The mean time between failure fell to its lowest point, close to 350 hours. Each computer had a personality, and the three in the aircraft were the most reliable and best behaved. A failure of one of them in the air led NASA to conclude that there was a systemic problem. Each computer was modified to fix prior problems, and they were at several different modification points, depending on their idiosyncratic failures and how long they could be spared from the development and verification effort. Jarvis and Szalai decided to halt flight research and send each one of the computers back to IBM for a complete refurbishment, to bring all of them to the same modification level, and to hard-wire some of the critical circuits. They sent a couple computers at a time back to Owego, New York, so that work could continue with the others. Four months later, in early January 1977, they were all back.[5]

---

[3] F-8 Digital Fly-By-Wire Flight Report, Flight 80-43-1, 27 Aug. 1976.
[4] F-8 Digital Fly-By-Wire Flight Report, Flight 81-44-2, 17 Sept. 1976.
[5] Kenneth Szalai, telephone interview, 30 Sept. 1998.

Computer serial number three was in its old Channel A position as Gary Krier took off on 28 January 1977, intending to accomplish some of the lost mission objectives from flight two. Thirty-eight minutes into the flight, at Mach 1.1 and 40,000 feet, almost the exact conditions of the previous failure, the Channel A fail lights lit up again. Krier turned the aircraft toward the runway and landed once again using the two good channels. The self-test routine detected an error in memory. The program tried to restore operation, trying 19 restarts before giving up and declaring a self-fail. Immediately after the flight, the engineers sent the computer back to IBM for yet another refurbishment.

On the one hand, the back-to-back failures of a computer in flight were frustrating, especially after grounding the aircraft and losing months of time to seemingly no avail. On the other hand, the system handled the failures well. There was an uneventful reconfiguration after these first two in-flight failures; less than 300 milliseconds elapsed for identification; no unexpected movements of the control surfaces occurred; no change in the flight-control system performance was noted by the pilot. Nevertheless, IBM's projection for mean time between failure after the refurbishment was 1,030 hours. The actual figure for the first five machines was 354 hours by late April 1977, almost the same as in the previous fall before they were sent back to Owego.[6] This dismal record was monitored by the increasingly anxious Space Shuttle engineers. Jarvis distributed copies of the flight reports to the Johnson Space Center and Rockwell.

One positive result of the flight was the performance of the uprated engine that the crew installed during the stand-down. Krier said that it turned the airplane into a "hot rod"—one that performed almost as well as the F-104. He could understand why LTV could still market the F-8 after 20 years in service.[7]

Krier flew two flights in February 1977 and one in early March without incident. These tested the autopilot and the augmented modes. There was mixing of the modes such as takeoffs with pitch-and-roll stability augmentation on and yaw in the direct mode. When yaw was also switched to the stability-augmentation mode, Krier had some jerking when the ailerons engaged with the rudder for turns, indicating a problem with the interconnection. Thomas McMurtry flew his first Phase II flight on 14 March, praising the handling qualities of the aircraft. These four successful flights provided increasing confidence in the system. The frequency of flights was about once every two weeks after they resumed in January. Now followed a burst of eight flights in a little over four weeks in direct support of the Shuttle Approach-and-Landing Tests, which were to begin in the summer.

---

[6] Calvin Jarvis notebook, 4/77-3/78: 20 Apr. 1977.
[7] Wilson Diary, 1975-1977, 28 Jan. 1977.

The First Space Shuttle Support Flights

Draper Laboratory built the software for the Space Shuttle Backup Flight System. The original idea was to use five computers in the primary control system with no backup. When all the actuators, cable runs, and digital channels were sorted out, the Shuttle engineers decided to use only four of the computers. The fifth computer, since there already was space and power, was kept as a backup. NASA contracted with International Business Machines Corporation (IBM) for the software in the primary system. When the fifth computer dropped out of that configuration, NASA decided to obtain the backup software from a different vendor.[8] Rockwell won that contract and turned to Draper Lab for the work, the reasoning being that a different set of programmers would help reduce the probability of a generic software problem.

Final approach and landing was a critical mission phase for the Shuttle, more critical than on a conventional aircraft since there were no engines for it. This meant one try, and one only, at making a successful landing. The Soviet Union, building its Shuttle ten years later, kept air-breathing engines in the design for help on approach, not wishing to take the risk the Americans took. To practice landings at minimum cost, NASA bought a used Boeing 747 from American Airlines, modified the tail to make room for the Shuttle, and used it to launch the Enterprise. This was a Shuttle Orbiter minus a number of systems not needed for approaches and landings, such as thermal protection tiles, engines, reaction controls, and orbital maneuvering systems.

The F-8's eight support flights in the spring of 1977 carried the Shuttle Backup Flight System's software test package. With it running in parallel with the F-8 flight-control software, the pilot would enter code 60 on the Computer Interface Panel to shut off the aircraft's usual downlink data stream and switch to the piggybacked software's downlink. He would make a series of Shuttle landing profiles; then he would enter code 61 to switch back to the normal downlink.

Ken Szalai ran the Shuttle software in the simulator when it arrived. One of the preflight test programs had a restart occur during execution. Szalai, very concerned, visually inspected the listing that Draper Lab always sent with the flight tapes against one made at Dryden. He was surprised to find that two locations on the tape did not match those on the listing, despite several layers of checking and cross-checking during tape manufacture. The first Shuttle support flight was only a couple days away, so he immediately began work on a fix. He got the program to work properly by manually entering the code from the listing onto the tape. However, according to the usual procedures, any fix would have to be made and re-verified at Draper Lab, which would take days. The software managers agreed that they would

---

[8] Tomayko, *Computers in Spaceflight*, Chapter Four.

just skip the preflight test involved and work around it. At a briefing on the problem on 17 March 1977, Szalai asserted that there would be no compromise in flight safety because the test simply checked values in rate gyros and accelerometers, and there were other tests that verified sensor integrity but did so indirectly. Also, the sensor redundancy management program provided safeguards as well. Szalai recommended the flight take place and that a long-term solution be found. He needed Draper Lab to concur. Vincent A. Megna faxed Szalai one paragraph of agreement the next day, almost as the aircraft was taxiing out to the runway.[9]

McMurtry then flew the first of the Shuttle support test flights on 18 March 1977. He rehearsed the free flight profile number one to lakebed runway 17. The Shuttle itself made approaches at a much higher speed than a powered aircraft. To simulate this, McMurtry kept the power pulled back and deployed the speed brake. He dropped at a very high descent rate six times toward the desert, each approach consistent with the others.[10] Krier flew profile number two on 21 March, also with good results. The next day McMurtry went up in the morning to practice profile five, and Krier flew profile four in the afternoon. Three weeks later, the pilots switched order. Krier tried out profile two again and McMurtry profile one in the afternoon. The next day, 15 April, this set of test flights ended with another doubleheader: McMurtry profile two and Krier profile five. Now the F-8 program had assisted the Shuttle development with aerial work to complement the experience of solving hardware and software problems with the AP-101. Flights of the F-8 halted then for two months while Langley Research Center's Remotely Augmented Vehicle experiment went through final preparations.

The Remotely Augmented Vehicle

Since the beginning of planning for Phase II, Langley worked to make its contribution to the program an exploration of advanced control laws. To make it easier to change the software containing the control laws, the Flight Research Center proposed having it resident on a ground-based computer. This would avoid the costs of verifying complex research flight-control laws. Telemetry downlinks would provide the vehicle state to the computer on the ground; it would do its calculations and then uplink commands to the actuators, just as though the machine and its software were on the airplane. This method has obvious risks, so, even though there would still be two flight control systems on the aircraft in case of telemetry or computer problems, Langley opposed using a ground-based system. Weeks of discussion followed, but the issue created an impasse between the two NASA centers. It

[9] Kenneth Szalai, "The Bug in the Tape Problem," briefing slides, 17 Mar. 1977, Dryden History Office; Vincent A. Megna, memo to Szalai, 18 Mar. 1977.
[10] F-8 Digital Fly-By-Wire Flight Report, Flight 87-50-8, 18 Mar. 1977.

seemed that Headquarters might have to make this decision.[11] But by spring 1974, everyone finally agreed on using a ground-based computer. During the aircraft modification, space then had to be found in it for an antenna, digital-to-analog converter, and a receiver/decoder. Robert Borek and James Craft identified places by early 1975.[12]

By the final design review for Phase II in late May 1975, Langley reported that it had finished advanced control law concept one, that for a control configured vehicle. (This was encapsulated in software alone; Dryden abandoned the plans to physically modify the F-8.) Langley's engineers developed an integrated package including the direct and augmented modes. The design objectives included aircraft-maneuver load control and gust-load alleviation using direct lift plus envelope limiting, with no modification to the F-8 aerodynamic characteristics. This set of control laws would thus demonstrate many of the advantages of control configured vehicles: smoother ride in turbulent air and increased lift by using all surfaces to develop it. Horizontal stabilizers conventionally produced downward forces for balancing an aircraft about its center of gravity. This configuration would obviate the need for that by maintaining active stability. The Langley engineers completed a version of the software with this functionality coded in FORTRAN by February, and it worked well in the simulator. The arrival of an AP-101 at Langley would allow them to complete their work.[13]

The ground-based control system initially consisted of a simplified version of the roll-and-yaw stability augmentation and pitch-control augmentation modes. There was no autopilot or side-stick support. Structurally, the software had an executive routine that contained the interrupt structure and the synchronization logic, plus five subroutines. Four of them executed the control laws, one of which handled the trim commands in a faster inner loop, with general feedback in a slower outer loop. The other one performed initialization and ran synchronization in the background. The telemetry downlink data went directly into the routines, and the executive controlled the uplink of the four 10-bit command words.[14] The remote-augmentation experiment started with a sample rate of 100 per second, but this could be easily adjusted.[15] The pilot would engage the ground system by entering code 21 for pitch, 22 for roll, and 23 for yaw via the Computer Interface Panel. Shortcuts included code 24 for both roll and yaw, and 25 for all three. Voluntary disengagement was through selecting a mode other than that in the remotely augmented vehicle experiment list.[16] The system would automati-

---

[11] James Phelps, log number 4, 25 Jan. 1974; 11 Feb. 1974.
[12] James Phelps, log number 5, 4 Feb. 1975.
[13] Calvin Jarvis, memo for distribution, 4 June 1975.
[14] Kevin Petersen, "F-8 RAV Baseline Experiment," S75-15-014, 31 Oct. 1975.
[15] Kenneth Szalai, Phillip Felleman, and Joseph Gera, "Design and Test Experience with a Triply Redundant Digital Fly-by-Wire Control System," AIAA Paper 76-1911, delivered at the AIAA Guidance and Control Conference, San Diego, CA, 16-18 Aug. 1976.
[16] NASA Fact Sheet FS-802-84-1, "F-8 Digital Fly-by-Wire," 24 Jan. 1984, p. 30A-32.

cally disengage because of uplink failure, a surface command rate exceeding a preset limit, or the autopilot becoming engaged. The machine used in these experiments was a Varian V-73 engineering minicomputer. For the remotely augmented vehicle flights, the AP-101s contained software with reasonability constraints and did nothing else except check the uplink signal for sensibility before passing it on to the actuators. Interestingly, the aircraft envelope for using this mode had a minimum value instead of a maximum: 15,000 feet or above so a signal could be received without any ground interference.[17]

It was difficult to start the remotely augmented vehicle tests due to failures and aborted flights. Flight 16 of the Phase II program took place on 15 June 1977, delayed a month due to a modification to the mode control panel to eliminate mechanical switching problems. Digital caution lights came on in the air without any related warnings from the annunciator panel, so the controllers aborted the flight and Krier returned to base. The remotely augmented vehicle mode was activated in a monitor setting. The crew then observed a couple of software problems that were quickly corrected.[18] Two weeks later, the next flight had a goal of open-loop checkout of the remotely augmented vehicle mode. A low-engine-pressure warning caused this flight to be aborted as well.[19] Finally, Tom McMurtry flew an uninterrupted test on 15 July, reporting that it was difficult to stop rolls at the desired attitude in the remote-augmentation mode.[20] That was the last open-loop preliminary test of the system because flight 19 suffered a computer failure four minutes after takeoff. Channel A again was the culprit, except that AP-101 serial number four was in place there instead of the twice-failed number three. Ironically, four was a replacement for serial number two, which went out during the preflight tests.[21]

As these initial shakedown flights took place, the Dryden Systems Analysis Branch wrote the final version of the closed-loop remote augmentation software. Kevin Petersen led this effort. There was a design review of this software on 31 May 1977, and only a little more than two months later a software readiness review for flight took place. Petersen named this version of the program RAVEN (Remotely Augmented Vehicle Experimental Norm). It was the baseline for later remote-augmentation experiments.[22]

RAVEN had its first test on 8 September 1977. Krier tried out all the flight axes individually and then collectively, noting that the software engaged and disengaged as planned with no transients. He repeated the tests six days later. After that, there was a break of over four months while the project tried some tests of ride-smoothing software and recovered from yet another

---

[17] Kevin Petersen and Kenneth Szalai, "F-8 Software Release Documentation," F8-77-014, 3 Aug. 1977.
[18] F-8 Digital Fly-By-Wire Flight Report, Flight 95-58-16, 15 June 1977.
[19] F-8 Digital Fly-By-Wire Flight Report, Flight 96-59-17, 29 June 1977.
[20] F-8 Digital Fly-By-Wire Flight Report, Flight 97-60-18, 15 July 1977.
[21] F-8 Digital Fly-By-Wire Flight Report, Flight 98-61-19, 20 July 1977.
[22] "F-8 DFBW RAV Software Design Review," 31 May 77, briefing slides; "F-8 DFBW RAV Software Readiness Review," 5 Aug. 1977, briefing slides, Dryden Flight Research Center History Office.

computer delay. By early September, there were no spare computers for flight test.[23] When McMurtry suffered a hard failure of Channel B (computer number five) on 15 September, there followed two months on the ground while IBM fixed enough computers to provide some margin. On 18 January 1978, Krier tried the remote-augmentation again but had to quit because the antenna failed. McMurtry was able to accomplish the evaluation on 14 February, reporting that lateral response was too sensitive with the remote system engaged. That spring, several additional flights expanded the envelope of the RAVEN system.

One of the chief advantages of using RAVEN was the low cost and time involved in making changes because of the high-level programming language and ability to have the experimenter instead of an outside group change the program. In addition, the flight verification testing was considerably less than with a new program on a computer aboard the aircraft—resulting in considerable time and cost savings. Costs were $10-20 per word and one-day turnaround for changes to the remote-augmentation software versus $100-300 per word and two weeks or up if Draper Lab did the work on the embedded system. The high point of 1978 for the remotely augmented vehicle system was a series of 15 flights, seven to explore low sample rates, eight for testing adaptive control laws. The sample rate record was 6.7 samples per second, about one-sixteenth of the initial rate.[24] From 2 February to 9 March 1982, a series of six flights using the remote-augmentation technology showed its continued flexibility. These flights tried out the concept of variable gains. This experiment was by a team from Dryden implementing the ideas of a group from the Royal Aircraft Establishment in England. Called the Cooperative Advanced Digital Research Experiment (CADRE), it had a baseline loop with fixed gains and another loop in which the gains dynamically varied based on actual performance in near-real-time.[25] There was a follow-on series of 24 advanced CADRE flights during the year from June 1982 to May 1983. The entire remote augmentation test program was successful enough to merit copying by the Advanced Fighter Technology Integration (AFTI)/F-16 project for some of its work and to be considered by others.

A Second Round of Shuttle Support

The free-flight portion of the Shuttle approach and landing tests began on 12 August 1977.[26] At the end of the fifth flight, 26 October 1977, the Enter-

---

[23] F-8 Digital Fly-By-Wire Flight Report, Flight 99-62-20, 8 Sept. 1977.
[24] "Remotely Augmented Vehicle Experiments," briefing slides, Dryden Flight Research Center History Office, 18 Oct. 1978.
[25] Richard R. Larson, Rogers E. Smith, and Keith Krambeer, "Flight Test Results Using Nonlinear Control with the F-8C Digital Fly-By-Wire Aircraft," AIAA paper 83-2174, delivered at the AIAA Guidance and Control Conference, Gatlinburg, TN, 15-17 Aug. 1983.
[26] They were preceded by captive-inactive and captive-active tests; see Richard P. Hallion, *On the Frontier: Flight Research at Dryden, 1946-1981* (Washington, DC: NASA SP-4303, 1984), pp. 357-358.

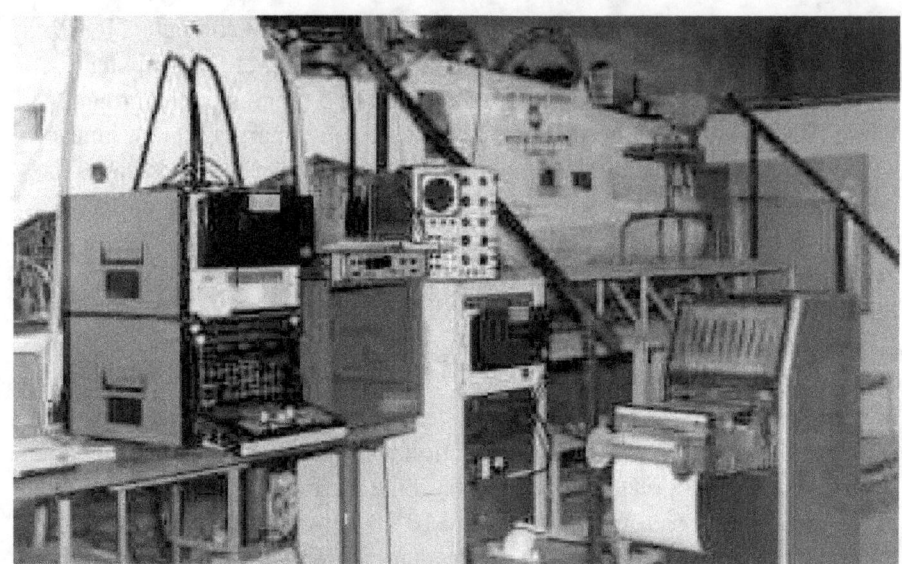

The heavily instrumented Iron Bird with three AP-101s in the avionics bay. The Shuttle support missions were practiced over and over, using the simulator as the primary tool. (NASA photo ECN-5220).

prise suffered a major pilot induced oscillation. Fred Haise of Apollo 13 fame was at the controls and Prince Charles of England in attendance as the big blocky glider quickly descended in perfect weather. As it closed to within thirty feet of the runway, Haise rolled slightly, seeming to search with the main gear for solid ground. A moment later he touched down hard and the Shuttle bounced, pulsed down in pitch and rolled sickeningly to the right. The roll kept up for a few cycles until the craft expended enough energy to make a landing unavoidable. This oscillation seemed to happen because of transport delays in the control system. Between the time the pilot moved the control stick and the time something actually happened at the control surface, there was a gap on the order of 200-300 milliseconds. The delay was due to the time needed to do analog-to-digital signal conversion, control law execution, and digital-to-analog conversion, as well as to the length of the wires and lag in the hydraulics. If the delay is too long, the pilot will lose patience and deflect the control device even more, but by that time the first set of commands is in process, so the effect is soon much higher, causing an overshoot. Seeing this, the pilot reacts by giving an opposite command, and it may result in an overshoot in the other direction. Then the task becomes one of damping the oscillation while the runway quickly grows outside the window. The pilot also must be in a "high gain" situation, such as landing or tight tracking. Otherwise, the PIO is not likely to show up. That was why it was not picked up on the thousands of hours of Shuttle simulation.

The Shuttle program asked the F-8 team to help find out the range of transport delays within which the pilots are likely to avoid inducing oscillations. The engineers first tried to do this using the remote-augmentation system. Tom McMurtry and Gary Krier each tried one flight evaluating various delays and trying high-speed approaches. To help cut drag and keep the kinetic energy up, the mechanics removed the landing gear doors and the pilot kept the wing in the down position,[27] resulting in approach speeds of over 200 knots, closer to that of the Shuttle. The flights occurred on 24 and 25 March 1978, and both pilots lost the radio connection to the ground facility due to excessive vibration and blown fuses. The tests would have to wait until an on-board version of the transport delay software arrived.

While the project waited for the software upgrades, two more pilots checked out in the F-8. Einar K. Enevoldson and John A. Manke came to the project bringing experience landing lifting bodies, which exhibited low lift-over-drag profiles like those of the Shuttle. In fact, Manke held the record for most flights in the lifting-body programs.[28] Both flew high-speed approaches as part of their check rides. Now four pilots could help attack the oscillation problem.

Draper Lab built variable delays into the flight-control software, and these could be selected through the computer interface panel and the five-position switches. On 7 April both McMurtry and Krier tried out the new program, flying seven and eight approaches respectively on one flight apiece. McMurtry commented that when the transport delays got too long or gains were wrong, he found himself adapting to them by pulsing the controls. He commented in his pilot report, "It was almost as if I was playing some kind of game with the airplane. I would make a gain change and then I would see how I could adapt to that gain to make the airplane behave reasonably well. I was surprised to [see] how well I could adapt."[29]

Six more flights followed within the next ten days with all four pilots participating. Counting the first two flights on 7 April, they flew 57 high-speed approaches. The Enterprise would have taken over a year to do the same since it was limited to one approach per flight and each flight required loading on to the 747. On the 18th, John Manke took off, intending to add to the total of high speed approaches. On the last of his six that day, he roared to the runway at 265 knots with 100 milliseconds of delay added to the F-8's usual response time. From the ground camera film of that approach, it looked like Manke pulled the nose up a little too high on the ensuing takeoff and compensated with a quick and clearly excessive downward pulse, causing the F-8 nearly to land nose-first. It took over five pulses to settle on a good

---

[27] The F-8 had a variable incidence wing with a hinge so that pilots could raise the wing to a higher angle of incidence for low-speed landings on carriers. To simulate the Shuttle landing speeds, pilots did their approaches in the F-8 DFBW with the wing at the lower incidence, what is referred to here as the "down position."

[28] R. Dale Reed with Darlene Lister, *Wingless Flight: The Lifting Body Story* (Washington, DC: NASA SP-4220, 1997), p. xviii. Actually, Milt Thompson had a higher number of total flights, but almost all of them were in the lightweight, unpowered M2-F1. Manke had more flights in the powered lifting bodies than anyone else.

[29] F-8 Digital Fly-By-Wire Flight Report, Flight 117-80-38, 7 Apr. 1978.

departure attitude and begin to gain altitude. The film from the tail camera is quite dramatic, with the runway filling the screen on the first oscillation.[30] Actually, what happened was that the control system worked against him, lengthening the recovery. He reported trouble in controlling pitch the entire flight. On the fifth approach, he commented, "[I] just got it in my mind I was going to drive that thing into the ground." Manke used a series of small pitch inputs to get there, "probably a classic PIO [pilot induced oscillation] down to the ground." As he rotated after the high-speed touch-and-go, the landing gear struts suddenly compressed, causing the pitch rate gyros to respond in such a way that the software disengaged the control/augmentation system and put it in the direct mode. The direct mode had degraded handling qualities relative to the control/augmentation mode, making it more difficult to recover. Manke first turned off the delay, then manually engaged the pitch stability/augmentation system to help in finally damping the oscillations.[31] As a relieved Manke pulled away, Gary Krier deadpanned from the control center, "Uh, John, uh, I don't think we got the data on that. We'd like to have you run that one again."[32]

This series of flights produced valuable data about handling characteristics with transport delays from 20 to 200 milliseconds. By flying the Shuttle landing profile, the pilots gathered data that helped set reasonable sample rate and control law execution limits. These tests also resulted in the Dryden-developed PIO-suppression filter that was tested in the F-8 and implemented into the Shuttle software prior to the first flight, eliminating the problem discovered by Fred Haise. This ended direct Shuttle support, though research on pilot induced oscillations occupied the program on over 30 more flights. The F-8 team moved into what was informally known as "Phase IIB," a series of experiments by Dryden engineers and those from outside the Center. The CADRE experiment (treated above) is one of these. Three of the others that had significant impact were: adaptive control laws, sensor analytic-redundancy management, and resident backup software.

Adaptive Control Laws

The teams at Langley and Dryden agreed to develop two control law packages for Phase II. The baseline was the Active Control Law Set. This demonstrated the potential benefits of active control for future aerospace vehicles by using algorithms that went beyond conventional stability augmentation systems. Aside from improved handling, gust alleviation, and drag reduction, the active control laws could control the aircraft while unstable in pitch due to moving the center of gravity aft. This reduced-static-stability

---

[30] Videotape No. 30, "Aeronautical Flight Research," in the NASA Dryden Historical Video Reference Collection.
[31] F-8 Digital Fly-By-Wire Flight Report, Flight 125-88-46, 18 Apr. 1978.
[32] Kenneth J. Szalai, interview with Lane Wallace, Dryden Flight Research Center, 30 Aug. 1995.

configuration is used in the F-16 and many later fly-by-wire aircraft. The F-8 team based the active control laws on a design study by Honeywell, Inc. that explored the linear quadratic optimal control method, which first finds the best weighting parameters, then finds the best feedback gains, and finally makes these into gain schedules that achieve the performance criteria.[33]

When the conversion of the F-8 from Phase I to Phase II just began, the Honeywell engineers made a proposal for the second type of control laws, the Adaptive Control Law Set. They suggested an experiment to try out adaptive control laws on the simulator at Langley, then in flight. Eventually the adaptive-control-laws experiment had the broadest base of any in the F-8 program, involving researchers from universities, industry, and government. The original idea was to use the power of digital computers to make the control laws able to adapt to a changing flight environment, improving handling and reliability. Adaptive control laws had been a dream of airplane designers for decades. Aircraft operate in a constantly changing aerodynamic environment. Dynamic pressure, Mach number, angle of attack, etc., cause wide variations in air data. The configuration of the aircraft also causes variables: the location of internal and external stores, such as fuel tanks and weapons, as well as the geometry of the airplane. In an analog system the control laws are necessarily fixed in hardware. In the Apollo computer during Phase I, no matter what happened, the control laws in software executed statically (in each mode such as wing up or wing down). Changes to them meant re-programming. Adaptive control laws would take outside data and the results of previous commands and dynamically project the best solution.[34]

Honeywell engineers initially proposed up to five candidate methods to create adaptive control laws. These were the Newton-Raphson technique for identification and algebraic gain solutions, a recursive least-squares approach, model tracking, high-gain model following, and the Lyapunov model reference technique. The people from Honeywell thought NASA preferred the first two, but they gained expertise in all five through various study and development contracts. The common thread in them all was high mathematics volume, which could only be accommodated on a powerful digital computer. Langley sponsored projects to explore some of these candidates. R.C. Montgomery of that research center teamed with two City College of the City University of New York professors, R. Mekel and S. Nachmias, to define a learning system for control. It had three components. The first was an information acquisition subsystem that used either the Newton-Raphson technique or Lyapunov's method to identify instantaneous values of the

---

[33] S. R. Brown, R. R. Larson, K. J. Szalai, and R. J. Wilson, "F-8 Digital Fly-By-Wire Active Control Law Development and Flight Test Results," manuscript, 26 Aug. 1977, in the Dryden Flight Research Center History Office.

[34] Honeywell, "Honeywell Digital Adaptive Control Laws for the F-8," Volume 1, Technical, Minneapolis, MN, 15 Jan. 1974.

parameters and then monitor their performance. The second component was a learning algorithm subsystem that related the actual performance to a model of projected performance using the least-squares technique. The last component computed the feedforward and feedback gains.[35] Dr. G. Kaufman of Rensselaer Polytechnic Institute studied the model-following technique. Dr. M. Athans of MIT suggested a multiple modeling system.

In mid-1977, two Honeywell engineers proposed another method. Gary L. Hartmann and Gunter Stein developed what they called Parallel Channel Maximum Likelihood Estimation (PCMLE) software. This program calculated the most likely gains in the longitudinal axis by measuring pitch rate, acceleration, and stabilizer position. Honeywell implemented it in FORTRAN-IV on a CYBER computer using several parallel channels of Kalman filters. The program had likelihood functions for each channel, and the gains from the channel with the maximum likelihood were used for controlling the aircraft.[36]

A selection committee consisting of Jarvis, Szalai, and Deets of Dryden and John Bird, Jarrell Elliot, Raymond Montgomery, and Joseph Gera of Langley decided among the various schemes. The decision criteria were: Does the concept represent state-of-the-art thinking in the control or estimation field? Is there a potential payoff in the application of the concept to future aerospace vehicle control? Is there a sufficient theoretical basis? Will flight testing enhance the concept? The idea that seemed to have the best answer to all of these questions was PCMLE. They decided to try out three-, five-, and ten-channel implementations, and also a reduced-sample-rate version. Szalai's software team installed the final program on the Varian ground computer and built a program called RAVADAPT for the AP-101 computers that took care of the interfaces between the PCMLE system and the aircraft. They used actual flight data as inputs to the system during development, which helped tune the Kalman filters before flight.[37]

Gary Krier piloted all seven flights of the adaptive-control-law experiment. On 24 and 25 October 1978, he flew the flight profile planned for the experiment with the ground system in the monitor mode. Then in November there were five flights trying out the various channel and sample-rate combinations. The researchers learned that a low rate of 17.8 samples per second resulted in better following due to its scale being closer to the actual computer word size, and thus more accurate. They also found out that the five-channel system was superior to the three- or the ten-channel versions.[38]

One current application similar to the adaptive-flight-control concept is

---

[35] R. Mekel, R.C Montgomery, and S. Nachmias, "Learning Control System Studies for the F-8 DFBW Aircraft," manuscript, Dryden Flight Research Center History Office.

[36] Gary L. Hartmann and Gunter Stein, "F-8C Adaptive Control Law Refinement and Software Development," Honeywell Systems and Research Center, Minneapolis, MN, June 1977, pp. 9, 23, 27.

[37] B.A. Powers and K. J. Szalai, "Adaptive Control Flight Test Completion," Milestone Report F8-78-029, 17 Dec. 1978, pp. 1, 3-9.

[38] Powers and Szalai, "Adaptive Control Flight Test Completion," pp. 9, 13.

the use of tape cartridges on F-16C and D models to inform the flight-control computers of the aircraft configuration on any particular mission. The F-16A and B series had analog flight computers. A precursor system to adaptive laws, though exhibiting static behavior due to the hardware, was the gun tracking on the As and Bs. The cannon on the F-16 is mounted in the wing root on the left side of the fuselage. Since it is offset from the centerline, a special yaw-control circuit uses the rudder to compensate for recoil whenever

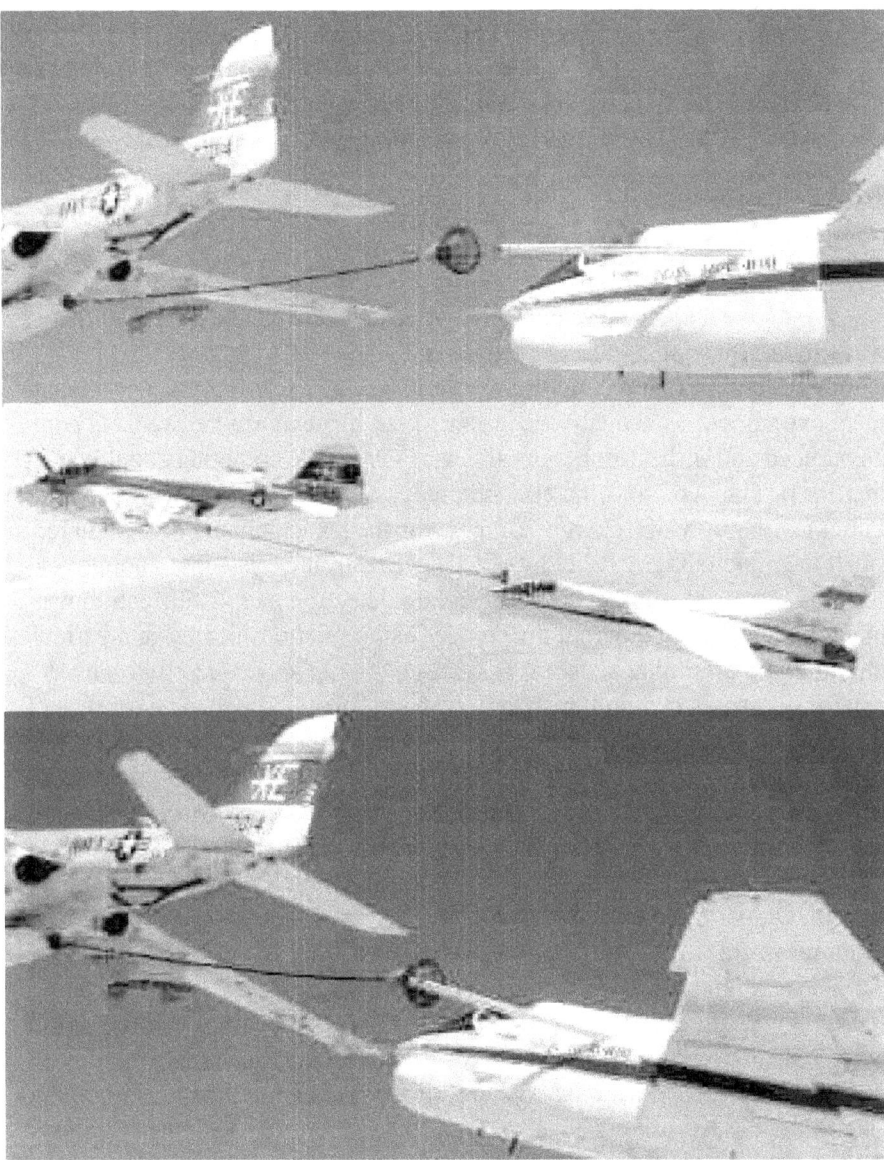

NASA research pilot Dick Gray flew a series of missions in 1982 with a dummy fuel probe to see how the fly-by-wire system handled in the aerial refueling environment. This three-photo sequence shows a successful link-up with a Navy KA-6 tanker. (NASA photos ECN-18439, ECN-18443, and ECN-18444).

the trigger is depressed. When digital computers came into use on the F-16, it became possible to tell them the location and weight of the fuel and weapons loads carried on external hard points. When a missile is fired, a fuel tank emptied, or a bomb dropped, the computers can maintain optimal performance by adapting their commands to the context of the new configuration.

The adaptive-control-laws experiment was Gary Krier's final one with the F-8 program. Stephen D. Ishmael, who joined the Center's cadre of pilots in 1977, checked out in the F-8 the day after the last adaptive-control-law test flight so the pool of project pilots could stay at the same level. In February 1979, Krier flew his last two missions to get vibration data on the flight pallet. He then moved on to law school and a second career in aerospace administration. The aircraft stood down until late September, as the project team prepared the next experiment, sensor-analytic-redundancy management.

Sensor-Analytic-Redundancy Management

The active and adaptive control laws received triplex-sensor data that was filtered by a redundancy-management algorithm, which yielded mid-values under non-failed conditions. The sensor-analytic-redundancy-management experiment used dual-sensor data to see if the number of sensors could be reduced and if the accuracy of the data sent to the computers could be increased. This work originated at Langley, which contracted with Draper Lab for it. James C. Deckert of Draper was the principal, and he assembled a small team including A. Willsky of MIT's Electronic Systems Laboratory.

The primary technique used for failure detection and identification was a method that evaluated output differences using relationships among variables measured by dissimilar sensor types. When the range between two sensor readings passed a certain threshold, one had probably failed. The essence of the problem was deciding which one. By using data from other sensors in certain arrangements, Deckert thought that he could figure out which sensor of the failed set was actually working. The second method was straightforward: constantly self-testing each instrument, and if its output over several cycles exceeds the physical limits of the aircraft, simply isolate it from the system. Draper Lab verified both methods by using actual data from previous flight research.

The analytic-redundancy-management algorithms used eleven dual sensor types: three accelerometers mounted in orthogonal axes, three sets of gyros, the angle-of-attack indicator, the altimeter, the Mach meter, and the vertical and directional gyros. The algorithm operated in a monitoring mode aboard the F-8; its presence had no influence on the sensor signals used by the control laws. That way, its results could be compared to actual events. Of course, it only used data from two of the triplex sensors. The flight code itself could simulate sensor failures. The software consisted of the analytic-redundancy-management section and the interface to the computer-input panel for the pilot to control the testing through selecting varied values for key parameters and failure injection of several kinds: drift, hardovers, loss of signal, and transient pulses. To stave off the combinatorial explosion of data made possible by using so many sensors, Deckert's team devised the sequential probability ratio test, which minimized the number of observations needed to choose between failures and non-failures.[39]

There were four types of analytic redundancy used in the algorithm. The rotational kinematics analysis related outputs of the rate gyros and the vertical and directional gyros. Altitude kinematics related the altitude given by the barometric altimeter and the double integral of the accelerometer and vertical gyro outputs. Translational kinematics involved the integrated outputs of the accelerometers, vertical gyros, rate gyros, and the air data sensors. Lastly, translational dynamics compared detected and calculated aerodynamic forces.[40]

The analytic-redundancy algorithm did not rush to judgment. If values differed from the previous sample by the magnitude of the failure threshold, the software declared the sensor provisionally failed. If the values returned to the expected range by the next sample-processing cycle or two, then the program lifted the provisional failure, or, if the values continued to be outside the acceptable range, it declared a hard failure.

This experiment started as early as June 1975 and ended with a final report in September 1982.[41] It began with a Langley study contract given to Draper Lab to explore the reduction in the number of control sensors. The initial work lasted two years and resulted in a FORTRAN program running on an Iron Bird at Langley Research Center, demonstrating the redundancy concepts. It took two more years of work to get the software ready for the actual aircraft. Finally, Steve Ishmael made the first flight test of the Phase I analytic-redundancy management software on 26 September 1979 (not to be confused with Phase I of the F-8 DFBW's overall flight research). He flew it again on 31 October, three more times between 21 November and 29 November, and then on 7 February 1980. Draper modified the software and refined the experiment for a Phase II flight-test program. Ishmael was the pilot on all of these as well. He made the first two flights in late October 1980 and March 1981. It became obvious that the directional gyro and vertical gyro outputs were difficult to analyze at high roll rates, so they were removed from the algorithm. The final three flights in June 1981 carried this modified software.

The experiment was successful in proving analytic redundancy could work. One frequently cited incident is that the analytic-redundancy-manage-

---

[39] James C. Deckert, "Definition of the F-8 DFBW Aircraft Control Sensor Analytic Redundancy Management Algorithm," R-1178, Charles Stark Draper Laboratory, Inc., Aug. 1978, pp. 1, 7, 19, 74-5.
[40] J. Deyst, J. Deckert, M. Desai, and A. Willsky, "Development and Testing of Advanced Redundancy Management Methods for the F-8 DFBW Aircraft," manuscript intended for the AIAA (unnumbered).
[41] James C. Deckert, "Analytic Redundancy Management Mechanization and Flight Data Analysis for the F-8 Digital Fly-By-Wire Aircraft Flight Control Sensors," Draper Laboratory, CSDL-R-1520, Sept. 1982.

ment software declared a sensor failed 24 seconds before the primary did. Even so, there is little use of this concept in operational aircraft today. One reason is that there is some comfort in hardware; the thinking is that despite the penalties, three sensors are better than two. Another reason is that the analytic-redundancy software was never flown as the primary source of sensor data, leaving one important step toward practical use not taken.

REBUS: REsident Back-Up Software

One of the last experiments flown on the F-8 digital fly-by-wire airplane was REBUS, or REsident Back-Up Software. Ken Szalai said that its story "shows the power of flight," because few believed in the idea until *after* the flight research series.[42] Szalai, Dwain Deets, and Wilt Lock were the Dryden engineers behind this effort, with Vince Megna leading at the Draper Laboratory. It spanned the years 1984 and 1985.

Since the primary flight-control system in the F-8 had three identical channels with identical software, there was a danger of a "common mode" or "generic" software failure bringing down the entire system with all three channels failing nearly simultaneously. Dissimilar software in the three channels would greatly reduce this possibility, but it would triple the expense of verification and validation, which had nearly broken the program testing only one implementation. This is why the aircraft had a Computer Bypass System as a backup and why the Shuttle program retained a fifth computer and loaded it with software with greatly reduced functionality that could do little else but bring the spacecraft home. The team wanted to see if the backup hardware could be removed from the design. The solution it employed was to use a relatively simple external hardware device to monitor fault declarations along with software resident in the computer memories that could provide basic flight control. The primary redundancy management system expected only single faults. If all three channels had alarms close together, the REBUS system assumed that there was a generic fault and forced a transfer to the software.[43]

The REBUS itself worked asynchronously and autonomously. It was a basic fixed-gain stability augmentation system. It could provide the aircraft with better handling qualities than the analog backup, but ones that were not as good as the primary system provided. There was no sensor redundancy management, no exchange of data, no synchronization, and no gain scheduling among the computers while running REBUS. If there was a hardware failure of one of the computers, there would continue to be output variances among the three machines. The system handled this by using mid-value logic on signals to the actuators. The largest obstacles to using the REBUS were

---

[42] Kenneth Szalai, interview, Dryden Flight Research Center, 11 June 1998.
[43] "Resident Backup Software (REBUS) Program," undated manuscript, Dryden History Office.

the initial switchover and the loose synchronization when it started running. Lock and Megna felt that they could reduce transients during switchover by using the last sensor data values sent to the primary system. Since the switching process took less than 200 milliseconds, the data would still be good. The synchronization process they defined was:

1. A computer set its own discretes high.[44]
2. It would search for high discretes from partners.
3. The computer would allow 400 microseconds for finding them.
4. If high discretes were found from at least one other computer, then the system would proceed or go back to step 1.
5. The computer then set the discretes low.
6. It searched for low discretes from each member whose high discretes were found.
7. The machine allowed up to 400 microseconds to find them.
8. The computer would exit the synchronization routine if synchronized with one other member; otherwise it would return to 1.[45]

After developing the software and the inevitable testing on the Iron Bird, the project team held a flight readiness review on 17 February 1984. The first research flight followed a few months later.

The three-channel control system flew for eight years prior to the start of the REBUS testing and was quite reliable. Therefore, instead of waiting for faults to appear, the REBUS software caused failures by writing to write-protected locations. The timing of the fault injection could be controlled by the pilot via the computer interface panel. Edward Schneider, a former Navy test pilot who joined the Center in 1983, took off on 23 July 1984 to evaluate the system using a very ambitious flight-test plan containing 81 items. The plan was to arm the REBUS at 0.6 Mach and 20,000 feet, check out the computer bypass system to be certain of a fallback, then transfer to REBUS. First Schneider would do pulses in all axes and some maneuvers. Then he would go back to the primary system to be sure that worked. He would then expand the envelope by going back to REBUS, doing some 2g maneuvers, downmode to the computer bypass system, back up to REBUS, back to the primary again, and repeat these cycles with different maneuvers a few more times. Finally, he would make simulated approaches in REBUS, do some touch and goes and low approaches, and then land, controlled by the primary system.[46] The test was so successful that Rogers Smith, a highly experienced research pilot who joined the Center in 1982, made the landing in REBUS after its second flight on the 27th. Schneider flew REBUS again in August,

---

[44] Discretes are values that signal an event or a state.
[45] Vincent A. Megna and Wilton R. Lock, "F-8 DFBW System Redundancy Management Analysis and Resident Backup Software," undated manuscript, Dryden History Office, pp. 8-12.
[46] F-8 DFBW Flight Card 231-194-152, Dryden Flight Research Center History Office.

Smith in September, and Schneider finished the test program in 1985 with flights on 1 February and 2 April. Problems with Channel A, weather, and a couple lateral-directional handling-qualities flights caused the gap from September to February. Most of February and March were devoted to a ten-flight series called the Optimum Trajectory Research Experiment (OPTRE) that involved data uplink and downlink between the F-8 and a computer in the new Remotely Piloted Vehicle Facility.[47] Schneider's last REBUS flight in April was the second to the last of the F-8.

In the six test flights, the REBUS was active for three hours and 54 minutes. There were 22 transfers from the primary system to it, with six of those at greater than one g. The experiment was a valuable contribution to the reconfigurable software systems developed for aircraft today and is used on the B-2.[48] It also focused the redundant versus dissimilar software debate, which eventually led to the differing implementations by Airbus and Boeing discussed in the next chapter.

Two F-8s in early retirement: the fly-by-wire and supercritical wing testbeds await refurbishment into permanent displays at the Dryden Flight Research Center. (NASA photo EC91-194-23 by Jim Ross).

---

[47] Wilton P. Lock, telephone interview with John "Dill" Hunley, Dryden Flight Research Center, 14 Oct. 1998.
[48] Calvin R. Jarvis, reviewer comment, Nov. 1998.

Denouement

After REBUS the F-8 was an airplane without a mission. There was a final tape release, 17C, which improved the aileron feel system. Schneider tried it out on 16 December 1985 with a flight aimed mostly at clearing the cobwebs and keeping the airplane flyable. The actual airframe was intended to be the basis of a new oblique wing test program, but it was cancelled due to a redirection of Navy fighter and attack aircraft tactics. (The Navy was a joint participant in the program.) So, the 169th flight of Phase II, the 211th of the overall program, was the F-8's last. After over 13 years of service as a digital fly-by-wire testbed, it sits gleaming in the desert sun on a pad in front of the Dryden Flight Research Center, the F-8 used for the Supercritical Wing on its right, the X-29 forward-swept wing demonstrator, built partly on its success, to its left.

Chapter Eight: The Impact and Legacy of NASA's Digital Fly-By-Wire Project

[Flight research is done] "to separate the real from the imagined."
—Hugh L. Dryden[1]

NASA spent over 15 years finding out what was real about digital fly-by-wire technology. The research program achieved important goals: it proved that active control could be accomplished with digital systems and that multiple computers could be synchronized and provide a greater measure of flight safety. It also demonstrated many other ideas, such as adaptive control laws, sensor analytic redundancy, and new methods of flight testing digital systems remotely. Dryden Flight Research Center proactively spread the word about the results in these research areas and helped stakeholders in the commercial application of the technology reach common ground on certification of digital systems. The digital fly-by-wire project contributed to the development of control technology and has a place in history there as well. The NASA experiments expanded the volume of engineering knowledge and made it applicable to other domains of control. Finally, the project also had a lasting impact on the Center itself through its alumni and the techniques they pioneered.

Airbus's adoption of fly-by-wire for its commercial transports is the ultimate proof that the objectives of the NASA program were met. In the 1980s, the A320 was the first commercial application, with others following in the 1990s. (Photo courtesy Airbus Industrie of North America, Inc.)

---

[1] Hugh L. Dryden, "General Background of the X-15 Research Airplane Project," in *Research-Airplane-Committee Report on Conference on the Progress of the X-15 Project* (Hampton, VA: Langley Aeronautical Laboratory, 25-26 Oct. 1956), xix.

Technology Transition

The F-8 project team members used several avenues to deliver information about fly-by-wire to the outside world. They also did not neglect internal technology transition. Ken Szalai held a series of in-house seminars on active flight-control and redundancy management in 1979, summarizing for others at Dryden what the new technologies were and how to use them. As noted above, the software development process acquired from the Draper Laboratory as early as Phase I of the program provided the foundation for later software efforts at the Center.

The F-8 program had a significant number of publications. The bibliography compiled for the 20th anniversary of the first flight lists 41 key papers directly reporting results from the project.[2] These papers first came out as early as the first flight research period and nearly continuously after that. The content and results of the Phase I flights were well documented in several papers given at a NASA-sponsored conference on fly-by-wire in 1974.[3]

Papers are passive in that one hopes people pick them up and read them. Workshops are a more active way for the results of the program to be transmitted, especially if key organizations are invited to send representatives. Dryden and Draper Lab offered a "Workshop on NASA Advanced Flight Control Systems Experience" at the Los Angeles Airport Hilton on 20 to 22 June 1978. The briefing slides show that they reviewed the feasibility aspects of Phase I, then the activities of Phase II. The purpose was to transition to aerospace corporations enough detail about the technology to encourage its application. At that time, there were already over 60 technical reports from the program, and the attendees received pointers to them for more information than the three-day briefing provided. This workshop was repeated in Cambridge, Massachusetts, at Draper Lab. This was the period that the F-18 development was underway, resulting in considerable industry interest in digital flight controls.

The presenters gave insights about flight software development, the newest and most daunting of the technologies. They reported that during the software construction phase, "transformation of design into code [was] more artistic than the hardware build." This is not what a nervous engineer considering digital fly-by-wire wants to hear, but such people were reassured by the fact that software defects discovered in post-flight analysis did not cause trouble in the air. Redundancy and self-testing provided shields from defects, as did the use of restarts for transient fault protection without even needing to know the source. There were only eight input/output errors in one billion operations by mid-1978, a statistic that also gave confidence to potential

---

[2] *Proceedings of the F-8 Digital Fly-By-Wire and Supercritical Wing First Flight's 20th Anniversary Celebration* (Edwards, CA: NASA Conference Publication 3256, 27 May 1992), Volume II.
[3] *Description and Flight Test Results of the NASA F-8 Digital Fly-By-Wire Control System* (Edwards, CA: NASA TN D-7843, Feb. 1975).

adopters of the technology.[4]

Some military programs immediately applied the lessons of the F-8 project. General Dynamics looked at both the F-4 and F-8 conversions into active control aircraft and gained enough confidence to make the F-16 fly-by-wire from its inception.[5] Lockheed's F-117A and Northrop's B-2 both could not fly unless active controls could be used, as their differing stealth technologies dictated unstable shapes: a flat-panel angular fuselage and a flying wing, respectively. All these projects are of 1970s origin and directly benefited from fly-by-wire research. The remotely augmented vehicle system is the ancestor of the Highly Maneuverable Aircraft Technology (HiMAT) flight-control system and all later remotely piloted vehicles. There are plans for unpiloted fighters to accompany piloted strike aircraft of similar design to provide less expensive escorts and bulk up attack groups without endangering humans. These aircraft would draw from the same control technology proved by the F-8 in the 1970s.

Dissimilar stealth technologies characterize the F-117A and the B-2, but fly-by-wire is needed by both. (U.S. Air Force photo).

The Certification of Commercial Fly-By-Wire Airliners

It was clear that the main stumbling block to the use of fly-by-wire in commercial transports was the certification of them by the U. S. Federal Aviation Administration and its international counterparts. Early in the 1980s, Boeing introduced two airliners with advanced avionics: the 757 and the 767. Even though the company had built a prototype cargo airplane using

---

[4] "Workshop on NASA Advanced Flight Control Systems Experience," briefing slides, Dryden Flight Research Center History Office.
[5] The early models of the F-18 used digital fly-by-wire, but the aircraft was statically stable and thus gained only limited benefits.

a fly-by-wire and hydromechanical system in the mid-1970s,[6] neither of the two new airliners had digital technology in their controls.

It was left to the relatively young Airbus to make the leap into the future and challenge Boeing's domination of the narrow-body airliner market with an advanced airplane using a flight-control architecture different from the American prototypes. Conscious of potential resistance by certifying agencies, pilots, and the flying public, Airbus engineers devised a scheme that is both redundant and a reduced-functionality backup at the same time. The objective was to introduce as much diversity in the system as possible (and thus avoid the dreaded generic software defect that could bring down all the computers) yet provide functional redundancy.[7]

There are two separate control systems in Airbus fly-by-wire aircraft (which now include the models A318, A319, A320, A321, A330, and A340). These work together to provide highly optimized handling in the pitch and roll axes. (The yaw axis still uses a mechanical system.) One is called the ELAC (Elevator Aileron Computers) and the other the SEC (Spoiler Elevator Computers). Thomson-CSF built the ELAC system, using two different computers, one designed in Paris, the other in Toulouse, by different teams not in contact during development.[8] In addition, the SEC, built by SFENA, is triply redundant. In order to achieve even higher reliability, the software is written in different languages, such as assembler unique to the processor, PL/M, and Pascal.[9]

If one or the other system fails completely, the remaining one becomes the backup. It is quite possible to fly an Airbus using spoilers for ailerons, and also without the spoilers, but the elevator control is built into both systems because that is still the primary control surface for pitch, and any other arrangement would not do. This rather Byzantine system is claimed to have a reliability of about one failure in 10 trillion operations, the highest ever achieved.

Boeing finally built an airliner with fly-by-wire controls, the 777. The control system is more straightforward than that used by Airbus. It contains three "lanes" of three different computers each: an AMD 29050, a Motorola 68043, and an Intel 80486.[10] Boeing is reportedly discovering one little-anticipated problem with choosing fly-by-wire: all the computer manufacturers supplying the processors are trying to stop new fabrication of these older machines. Now it is faced with a choice of buying up sufficient numbers of

---

[6] Called YC-14s, two were built by Boeing for the U.S. Air Force. Intended for short-field takeoff and landing, the airframe was unstable due to an unusual placement of the engines well forward on the upper surface of the wing so that the exhaust could enhance lift. First flown on 9 Aug. 1976, it had a mechanical control system with a triplex digital stability augmentation system. (John K. Wimpress and Conrad F. Newberry, *The YC-14 STOL Prototype* [Reston, VA: American Institute of Aeronautics and Astronautics, 1998], pp. 42-46, give an account of the system and the evolution of its redundancy management.)
[7] Etienne Tarnowsky, Airbus, telephone interview, 4 Feb. 1998.
[8] Pierre Condom, "Systems for the Airbus A320—Innovation in all directions," *INTERAVIA*, 40 (1985): 353.
[9] Tarnowsky, interview.
[10] John Aplin, "Primary Flight Computers for the Boeing 777," manuscript, p. 4.

the remaining production to outfit all projected 777s, both new and those in need of repair, or of trying to re-host the software, which also means recertification.

Since software is so difficult to certify, NASA tried to share its development experiences to help airplane companies and government agencies begin to work together on the problem. In 1976, NASA's Ames Research Center co-hosted a conference with the Federal Aviation Administration (FAA) on certifying flight software, in which members of the F-8 team participated. Attendees advised the FAA not to get into the business of certifying the software itself, but rather whether the functionality is present.[11] That treats hardware and software components as equals. The F-8 team made other contributions to the FAA's understanding. When Earl Dunham of Langley Research Center conducted additional lightning tests in late 1977, the FAA, which was worried about lightning and fly-by-wire, was briefed in detail on the results.[12] Cal Jarvis noted after that meeting that the FAA's only experience with digital systems consisted of certain autopilots, which had a reliability requirement with no clear sense of how it would be proved. He found out that the FAA had 50 test pilots, only about 30 of whom knew anything about digital systems, and only about 10 of whom could read block diagrams. Jarvis sensed that perhaps the FAA inhibited advanced designs because of its outdated certification requirements and lack of training among reviewers.[13]

Since then, there has been an evolution of recommendations for software development and verification. The latest (since 1992) is DO178B, an international standard that certifying agencies can require, as did the FAA in Advisory Circular 20-115B. Essentially, DO178B contains guidelines, not requirements, that can result in a disciplined software development process if followed. The process is similar to that for the F-8 and Shuttle programs, but it does not appear to be more complete or to have any innovations. Unfortunately, DO178B is only meant to show "intent," rather than giving clear direction.[14] The requirements list indicates *what* to do, not *how* to do it. This means the interpretation of what has to be done versus what is suggested to be done is discovered in FAA certification hearings with commercial airplane companies. There precedent is set and argued.

The F-8 Digital Fly-By-Wire Project in the History of Technology

The story of the development of fly-by-wire lies in both the history of aeronautics and the history of computing, and it contributes to the overall history of technology. Early chapters of this book gave the context of fly-by-

---

[11] NASA, "Government/Industry Workshop on Methods for the Certification of Digital Flight Controls and Avionics" (Moffett Field, CA: TMX-73, Oct. 1976), p. 9.
[12] Calvin Jarvis notebooks, 20 Dec. 1977; 23 Jan. 1978.
[13] Calvin Jarvis notebooks, 14 Mar. 1978.
[14] Leslie A. Schad, "DO-178B: Software Considerations in Airborne Systems and Equipment Certification," manuscript, p. 4.

wire in aeronautics and computing. The common theme is the maturation of the ability to control. Aeronautical engineers employed computers in flight-control systems not because they represented a new technology and were "progress for progress' sake," but because they were part of a solution to the flight-control problem. The development of NASA's first digital flight-control system did have some aspects of the application of interesting new technology just because it was new, but the program rapidly matured into a technology demonstrator for the exploration of many new techniques.

Control was always a part of using machines. The development of automatic feedback control enabled dramatic achievements. Control can be defined as "purposive influence toward a pre-determined goal." Commonly a program, which is any pre-arranged information that guides subsequent behavior, is needed to guide the system in achieving the goal. In early feedback control systems, such as Edmund Lee's use of fantails on windmills beginning in 1745, the program was encapsulated in the hardware.[15] This is similar to how analog computers carry their programming. The invention in 1822 of a cam to control the lathing of military gunstocks automatically used the same technology as the differential analyzer (a mechanical analog computer) did a century later. More recently the program has resided in software.

The ideas about control matured in the last century and a half. For Max Weber, the study of control meant the study of bureaucracy. Other social scientists and politicians followed this line of research and application, including Karl Marx, Vladimir Lenin, and Joseph Stalin writing about and experimenting with controlled economies. Right after World War II the focus changed. Since "information processing and communication are inseparable components of the control function," the increasing variety and availability of computers caused them to become intimately entwined with the concept of control.[16] Their information-processing role is obvious, and Claude Shannon gave the theoretical basis for the communication theory that is most useful in feedback control.[17]

Norbert Wiener invented the term "cybernetics" in 1948 to represent control and communication in animals and machines. His vision of automatic factories displacing workers caused great unrest among those likely to be replaced. Technologists loved and laborers scorned the word "automation." In 1953, feedback control systems were in highly automated factories, and by 1959 a Texaco refinery and an Imperial Chemical Industries processing plant used digital control systems. Some envisioned a "second industrial revolution" due to automation.[18]

---

[15] James R. Beniger, *The Control Revolution* (Cambridge, MA: Harvard University Press, 1986), pp. 7-8, 39, 176.
[16] James R. Beniger, *The Control Revolution*, pp. 6, 8.
[17] Charles Susskind, *Understanding Technology* (Baltimore: Johns Hopkins University Press, 1973), pp. 55-59.
[18] James R. Bright, "The Development of Automation," in Melvin Kranzberg and Carroll W. Pursell, Jr., eds., *Technology in Western Civilization* (New York: Oxford University Press, 1967), pp. 635-655.

In the 1950s, digital computers could be used in the role of factory controllers simply because factories were generally large places with plenty of room for the mainframes of the day. Dedicated control of smaller components, such as individual machines, had to wait for computers to shrink in size and yet increase in power. The demands of aircraft and spacecraft designers helped drive these developments. Integrated circuit manufacturers increased capability, quality, and production when the Apollo and Minuteman missile programs bought almost the entire supply of chips for several years in the early 1960s. By the time the engineers of the F-8 project adopted the Apollo Guidance Computer, the technology was in place for small, powerful, digital machines suitable for flight control in both aircraft and spacecraft.

What These Engineers Knew and How They Knew It

Walter A. Vincenti, an aeronautical engineer turned historian of technology, wrote a series of essays published in *What Engineers Know and How They Know It*. The book examined developments in aeronautical engineering as a demonstration of how engineers gain knowledge.[19] People think of engineering practice as a fairly straightforward application of known facts derived from scientific principles and designs from previous similar types. For instance, any bridge has the functional characteristics of all bridges, and the engineer's skill and creativity is exercised handling the variables such as the needed length and type of traffic. The stories in the book show engineers in the nascent aeronautical field often proving and refining ideas by experimentation. Every bridge is the culmination of knowledge gained from thousands of years of building successful bridges. Every airplane is the culmination of less than a hundred years of building successful airplanes. There are very few bridges and relatively many airplanes built at the edge of engineering knowledge. The role of engineers practicing at the forefront is to gain facts and techniques for their successors' use. This requires a different view of engineering work.

The NASA engineers at the Dryden Flight Research Center actively seek out projects that expand aeronautical engineering knowledge. They either invent or adopt promising new technologies, explore them, derive lasting principles, and transition this new common knowledge to other engineers. The technique, no longer new, becomes part of the practicing engineer's toolkit.

Fly-by-wire technology for flight control was not entirely innovative when Melvin Burke decided his team would work on it in the late 1960s. However, it was hardly in wide use by flight-control engineers, and, therefore, a prime candidate for a project at Dryden. The Air Force was already

---

[19] Walter A. Vincenti, *What Engineers Know and How They Know It* (Baltimore: Johns Hopkins University Press, 1990).

exploring the use of analog computers as the basis of such systems, abandoning digital computers after some work early in the decade. Because of how funding was allocated, the ostensible focus of the Air Force effort was survivability: fly-by-wire as a means to an end outside the flight-control problem. NASA could take a broader focus on maturing the technology for its own sake, while still keeping the eventual goal of a control-configured vehicle in mind.

NASA actually doubled the impact of digital technology. Piloted and unpiloted spacecraft and launch vehicles used digital computers for all aspects of flight control. Commercial launch-service providers and satellite builders now routinely use the technology in their vehicles and spacecraft. Similarly, commercial and military aircraft manufacturers now employ digital fly-by-wire. It has reached the point where designers of new military airplanes would hardly consider any other flight-control system. Within a few years this will also be true of civilian aircraft intended for commercial use. From both the standpoint of expanding engineering knowledge and of NASA's mission, the F-8 Digital Fly-By-Wire project is a success.

The Technological Legacy

The AFTI/F-16. (Photo courtesy of Lockheed Martin).

One follow-on project began even before the F-8 retired. It was the Advanced Fighter Technology Integration (AFTI)/F-16. As the initial flights of the F-8 helped lead General Dynamics to an analog fly-by-wire system for the F-16A/B, flights of the AFTI/F-16 (at Edwards Air Force Base with Dryden participation) led the firm to a digital fly-by-wire control system. One interesting function added due to the capability of the digital system was speech recognition and the capability for automatic recovery if the pilot lost consciousness due to maneuvers.

Perhaps the closest successor to the F-8 in the role of avionics pioneer is the F/A-18 Systems Research Aircraft. NASA maintains a fleet of F-18s as chase aircraft, the replacements for the F-104s. Some researchers at Dryden had the bright idea of putting experiments on one of the F-18s to take advantage of its frequent flight time. When word got around to avionics manufacturers that such a capability was available, the airplane garnered so much business that it is used in the experiment mode full-time. It pioneered fly-by-light, the use of fiber optic connectors, and the replacement of heavy centralized hydraulic systems with smaller units like electro-hydrostatic or electro-mechanical actuators that were also less susceptible to electromagnetic interference, battle damage, and a high incidence of maintenance than what they replaced.[20]

When Center Historian John "Dill" Hunley asked Ken Szalai about important follow-on projects to the F-8, his reply took a different direction than flight control.[21] He highlighted the Digital Electronic Engine Controls (DEEC) and the Highly Integrated Digital Electronic Control (HIDEC) programs. They did for engine controls what the F-8 project did for flight controls. In fact, their impact was more rapid and greater than flight controls themselves. They borrowed from the F-8 project its control law algorithm structure, its computational requirements, and its redundancy management. Commercial airplanes without fly-by-wire flight controls had engines with digital controllers. The pervasiveness of this technology is nearly complete. The certifying agencies like it because of its elegance and inherent redundancy: no commercial turbofan-powered aircraft has fewer than two engines. There are also the High Angle-of-Attack Research Vehicle (HARV) that pioneered thrust vectoring and the F-15 ACTIVE (Advanced Controls for Integrated Vehicles) that continued the work. Szalai pointed out that digital controls are not much affected by sand, wind, dust, and humidity, making them ideal for military and commercial applications.

More generally, as alluded to elsewhere in this study, the F-8 Digital Fly-By-Wire program served as the springboard for digital fly-by-wire technology to be used in both military and commercial aircraft. As already noted, the concern in the early 1970s was that digital fly-by-wire was just too complex and unreliable for piloted aircraft. Dryden's program demonstrated that this was not the case. As a result, not only the second-generation F-16 flight-control system became digital, but the first-generation F/A-18s adopted a digital flight-control system. Later, commercial transports like Airbus and the Boeing 777 adopted digital flight control as well, as noted above in this chapter.

The Human Legacy

The F-8 program was a fertile source of leadership for the Dryden Fight Research Center. Taking a snapshot of where some of the key pilots and

---

[20] Lane E. Wallace, *Flights of Discovery* (Washington, DC: NASA SP-4309, 1996), 124-126.
[21] Kenneth Szalai, interview with John "Dill" Hunley, Dryden Fight Research Center, 21 July 1998.
[22] Frank Hughes, interview, Johnson Space Center, Houston, TX, 2 June 1983.

engineers were at the end of July 1998 shows a significant number of alumni in important positions. Ken Szalai was Director of the Center. Kevin Petersen was Deputy Director, soon to be named Acting Director and then Director following Szalai's retirement on 1 August. Jim Phelps had recently headed the safety office before retiring. Dwain Deets commuted back and forth to NASA Headquarters as the Manager of the Research and Technology Flight Research Program, which encompasses most of the projects at Dryden and proposals for the future. Gary Krier returned to Dryden in 1995 and by 1998 led the Aerospace Projects Office. Cal Jarvis retired from that same position and in 1998 lived nearby. Tom McMurtry had been the Director for Flight Operations until 27 July 1998, when he became the Center's Associate Director for Operations. The successor to Krier as project pilot, Steve Ishmael, was the X-33 Deputy Manager for Flight Test and Operation. Pilots Ed Schneider and Rogers Smith also had significant levels of responsibility, with Schneider becoming the Acting Chief of the Flight Crew Branch and Smith, the Acting Director of Flight Operations on 27 July. Wilt Lock and Joe Wilson were "engineers' engineers," albeit senior to most of the others at Dryden. They both seemed more comfortable with engines and electronics than politics and budgets.

What can be deduced from this impressive list? The most obvious common characteristic among almost all of these men while they worked on the F-8 is youth. They were mostly in their 20s and 30s, and, with a few exceptions, were playing a major role in a program for the first or second time. In many ways they resembled the teams that made up the Apollo Project: young engineers, often right out of school, with seemingly limitless confidence and energy. NASA people who worked on Apollo and stayed with the agency for twenty years or more hold the fondest memories for that frantic period and time of their life.[22] In interviews for this book, there often was a similar sense of nostalgia and accomplishment among the F-8 DFBW alumni. However, if for many of those immersed in Apollo the rest of their careers seemed anti-climatic, this is not true of the F-8 engineers. They went on to several more projects before many of them landed on the administrative floor of Center Building 4800, each with challenges comparable to those of the fly-by-wire program. The F-8 achieved a major milestone in the history of technology when it flew with a digital computer. This is like the first lunar landing: a culmination, a defining moment. The program did not pioneer fly-by-wire, and by the time the Phase II flights began, the YC-14, the F-18, and several projects in England and Germany were either in the air or soon to be with multiple digital-control systems. Therefore, Phase II was an opportunity to experiment, to refine, to convincingly prove digital fly-by-wire was practical for many applications. This made the project similar to others at Dryden that had no moment where an historical marker could be placed, but which resulted in reams of data and practical experience. These were the projects the members of the F-8 team peopled in the 1980s and early 1990s. These are the projects in which Lock and Wilson still toil, separating "the real from the imagined."

Appendix: DFBW F-8C Flight Logs

Note: The flights are numbered xxx-yyy-zzz, where xxx is the number of the flight since NASA acquired the aircraft, yyy is the number of flights since it was converted to fly-by-wire, and zzz is the number of the flight in Phase II. The aircraft had 37 general familiarization flights prior to conversion. The Phase I flight numbers do not contain the zzz element for obvious reasons.

Phase I

Date	Number	Pilot	Objective	Remarks
25 May 1972	38-1	Gary Krier	First flight	
8 Jun 1972	39-2	Gary Krier	Envelope expansion and evaluation of DFBW and BCS systems	
19 Jun 1972	40-3	Gary Krier	Envelope expansion and evaluation of DFBW and BCS systems	-Stick gearing evaluation
18 Aug 1972	41-4	Gary Krier	Evaluation of SAS modes and new stick gearing	
22 Aug 1972	42-5	Gary Krier	Evaluation of SAS mode and new stick gearing	-SAS gave smoother takeoff -Landing had marginal control in SAS modes
15 Sep 1972	43-6	Gary Krier	Evaluation of SAS modes and new stick gearing	-Observed delay in roll between command and response -Tape reader malfunction
21 Sep 1972	44-7	Tom McMurtry	Pilot checkout and evaluation of SAS modes and new stick gearing	-Trim indicators reported insensitive -Plane tended toward lateral and directional PIO
27 Oct 1972	45-8	Gary Krier	New K-START tape and BCS evaluation	
3 Nov 1972	46-9	Gary Krier	New K-START, CAS and RRC evaluation, BCS Stick Gearing	-"Problems" listed as weak radio and left main gear indicator

				Evaluation	inop -BCS pitch axis now has non-linear LVDT
6 Dec 1972		47-10	Tom McMurtry	Evaluation of SAS, CAS, and BCS	
13 Dec 1972		48-11	Gary Krier	Evaluation of SAS, CAS, and BCS	
19 Dec 1972		49-12	Gary Krier	Evaluation of SAS, CAS, and BCS	
10 Jan 1973		50-13	Tom McMurtry	Evaluation of SAS, CAS, and BCS	
30 Jan 1973		51-14	Gary Krier	Performance comparison with SCW, handling qualities in GCAs, and CAS evaluation	
13 Feb 1973		52-15	Gary Krier	Stability and control maneuvers	
16 Mar 1973		53-16	Gary Krier	Stability and control maneuvers and tracking	
23 Mar 1973		54-17	Tom McMurtry	Stability and control maneuvers and tracking	-Focus on pitch
29 Mar 1973		55-18	Gary Krier	Stability and control maneuvers and tracking	-Focus on pitch -More tracking tasks
6 Apr 1973		56-19	Tom McMurtry	Stability and control maneuvers and tracking	-Yaw problems occur with lateral inputs -Pilot notes that plane wanders in lateral axes -TM signals lost in flight
6 Apr 1973		57-20	Gary Krier	Stability and control maneuvers and tracking	-Problems with BCS self test -Problems with the primary system in roll
1 May 1973		58-21	Gary Krier	Stability and control maneuvers	-TM signals lost, terminating some

				maneuvers -Smooth landing -Damped lateral control
8 May 1973	59-22	Gary Krier	Low speed handling qualities	
29 May 1973	60-23	Gary Krier	Stability and control maneuvers	
4 Jun 1973	61-24	Tom McMurtry	Low speed handling qualities	
7 Jun 1973	62-25	Tom McMurtry	Low speed handling qualities	-Servo warning lights appear early in flight -Problems with lateral-directional characteristics of airplane -Oscillations prevent target airspeed maintenance -Pilot experiences downmoding problems
22 Jun 1973	63-26	Gary Krier	Stability and control maneuvers	-Airplane is reluctant to roll in and out of turns -Some oscillatory and deviation problems -Pilot states that lateral control was significantly improved
26 Jun 1973	64-27	Tom McMurtry	Stability and control maneuvers	-Obvious overshoots when making bank angle changes -High pilot workload -Yaw wandering and lateral sensitivity -Tracking problems
3 Jul 1973	65-28	Gary Krier	Stability and control maneuvers	-Detailed data collection and rating of maneuver performance

Date	Flight	Pilot	Purpose	Notes
10 Jul 1973	66-29	Gary Krier	Stability and control maneuvers	-Pilot seems to have trouble with trim in certain mode/gear gain combinations
24 Jul 1973	67-30	Tom McMurtry	Stability and control maneuvers	-Focus on longitudinal handling qualities
19 Sep 1973	68-31	Gary Krier	Evaluation of force sensing and side-stick	-First flight with a stick on the side of pilot as opposed to the center
25 Sep 1973	69-32	Gary Krier	Evaluation of force sensing and side-stick	-First appearance of the "Side Stick Mission Rules Document"
3 Oct 1973	70-33	Gary Krier	Evaluation of force sensing and side-stick	-Pilot prevented from performing some maneuvers due to wind limitations that appear on the "Mission Rules Document"
4 Oct 1973	71-34	Tom McMurtry	Evaluation of force sensing side-stick	-Winds again prevent certain maneuvers in flight and reduce the altitude in flight
12 Oct 1973	72-35	Tom McMurtry	Evaluation of force sensing side-stick	-Crew appears to grow accustomed to wind limitations, now given alternate plans according to wind conditions on flight instructions
17 Oct 1973	73-36	Tom McMurtry	Side-stick evaluation	
24 Oct 1973	74-37	Phil Oestricher	Pilot checkout	-Favorable report from pilot; he is impressed by the DFBW crew and the quality of the airplane
24 Oct 1973	75-38	Phil Oestricher	Pilot checkout	-Pilot performs various maneuvers and formation exercises

Date	Number	Pilot	Objective	Remarks
31 Oct 1973	76-39	Bill Dana	Pilot checkout	-Tracking exercises cause PIO
8 Nov 1973	77-40	Bill Dana	Pilot checkout	-Various general evaluations -Alterations made to "Mission Rules Document"
19 Nov 1973	78-41	Einar Enevoldson	Pilot checkout	-Pilot evaluates primary control system
27 Nov 1973	79-42	T. K. (Ken) Mattingly	Control system evaluation	-Overall evaluation of all systems and side-stick

Phase II

Date	Number	Pilot	Objective	Remarks
27 Aug 1976	80-43-1	Gary Krier	Evaluation of DFBW primary and bypass systems.	-CBS roll self test failed, causing a startover of all preflights.
17 Sep 1976	81-44-2	Gary Krier	Envelope expansion and flying qualities evaluation.	-Flight was aborted due to channel A computer failure, caused by CPU parity error, but the first 9 test points were completed.
28 Jan 1977	82-45-3	Gary Krier	Flight test using refurbished AP-101 computers.	-Precautionary landing made due to 1 computer failure; landing made on remaining 2.
16 Feb 1977	83-46-4	Gary Krier	Evaluation of digital augmentation and autopilot modes.	
25 Feb 1977	84-47-5	Gary Krier	Expansion of augmented and unaugmented envelope.	

Date	Flight No.	Pilot	Purpose	Notes
2 Mar 1977	85-48-6	Gary Krier	Additional envelope expansion and handling qualities evaluation.	
14 Mar 1977	86-49-7	Tom McMurtry	Pilot familiarization with Phase II system and autopilot evaluation.	Altitude hold misfunction due to an attitude gyro abnormality.
18 Mar 1977	87-50-8	Tom McMurtry	Evaluation of Shuttle back-up flight-control system software test package with F-8 flying ALT free flight profile #1.	
21 Mar 1977	88-51-9	Gary Krier	Evaluation of Shuttle back-up flight-control system software test package with F-8 flying ALT free flight profile #2.	
22 Mar 1977	89-52-10	Tom McMurtry	Evaluation of Shuttle back-up flight-control system software test package with F-8 flying ALT free flight profile #5.	
22 Mar 1977	90-53-11	Gary Krier	Evaluation of Shuttle back-up flight-control system software test package with F-8 flying ALT free flight profile #4.	-Pilot notes that pitch stick forces were unpleasantly high.
14 Apr 1977	91-54-12	Gary Krier	Evaluation of Shuttle experimental software package with ALT free flight profile #2.	-Flight was postponed twice because of failing roll axis computer by-pass self-test.
14 Apr 1977	92-55-13	Tom McMurtry	Evaluation of Shuttle experimental software package with ALT free flight profile #1.	
15 Apr 1977	93-56-14	Gary Krier	Evaluation of Shuttle experimental software package	

			with ALT free flight profile #2.	
15 Apr 1977	94-57-15	Tom McMurtry	Evaluation of Shuttle experimental software package with ALT free flight profile #5.	
15 Jun 1977	95-58-16	Gary Krier	Control system envelope expansion and open loop RAV checkout.	-Flight was originally scheduled for May 17th, but cancelled due to mode panel switching problems. -Problems with Mach and altitude indicators.
29 Jun 1977	96-59-17	Gary Krier	Control system envelope expansion and open loop RAV checkout.	-At end of testing, straight-in landing made due to "engine low oil pressure" light illumination.
15 Jul 1977	97-60-18	Tom McMurtry	Handling quality investigation of the F-8 modes at 20,000 ft with speed range of 300-400 KIAS, with additional open loop maneuvers.	-Additional RAV testing done after the flight.
20 Jul 1977	98-61-19	Gary Krier	Handling quality investigation of the F-8 modes at 20,000 ft and speed range of 300-400 KIAS, with additional open loop maneuvers.	-Flight termination due to computer failure indication shortly after takeoff. -Flight was originally scheduled for earlier date, but channel A failure caused abort.
8 Sep 1977	99-62-20	Gary Krier	Handling qualities evaluation of various DFBW control modes, evaluation of automatic angle-of-attack limiter system, 1st formation flight for Phase II, and engagement of RAV mode.	-Substantial AP-101 problems existing in last several flights; many still outstanding. -Flight was previously aborted due to failure in analog CBS system.

14 Sep 1977	100-63-21	Gary Krier	Evaluation of handling qualities and angle-of-attack limiter, energy maneuverability, and RAV engagements.	
15 Sep 1977	101-64-22	Tom McMurtry	Envelope expansion for SAS and CAS modes, handling qualities evaluation, and maneuver flap tests.	-Flight terminated early due to channel B AP-101 failure. By-pass system correctly "by-passed" bad computer. Problems were delegated to IBM for investigation.
18 Nov 1977	102-65-23	Gary Krier	Evaluation of ride smoothing system.	-Because of the objective, test occurred in afternoon with severe turbulence.
21 Nov 1977	103-66-24	Tom McMurtry	Evaluation of ride smoothing system.	-Moderate levels of turbulence.
23 Nov 1977	104-67-25	Gary Krier	Evaluation of ride smoothing system.	-Light turbulence.
30 Nov 1977	105-68-26	Gary Krier	Investigation of autopilot control modes and performance of stall approach maneuver.	-Excessive workload in trim SAS and CAS; pilot believed this was due to stick shaping and sensitivity problems.
2 Dec 1977	106-69-27	Gary Krier	Evaluation of operation of angle-of-attack limiting system, control augmentation software, and gathering data on longitudinal trim changes.	
18 Jan 1978	107-70-28	Gary Krier	Performance of series of lateral-directional pulses, and investigation of longitudinal stick characteristics for a	-Inoperative communication antenna.

			proposed new location for the primary pitch LVDT. Also evaluation of RAV mode, handling qualities in fine turns and formation flight, gunsight tracking.	
14 Feb 1978	108-71-29	Tom McMurtry	RAV evaluation, envelope expansion for ride smoothing, flap derivative series, formation flying evaluation, gunsight tracking, pitch command LVDT sensor evaluation.	-Lateral response seemed "too sensitive" to pilot. -New parabolic stick shaping in software release improved roll response around center stick.
17 Feb 1978	109-72-30	Gary Krier	RAV evaluation, ride smoothing evaluation, lateral-directional evaluation.	-RAV handling qualities were "inferior" to basic SAS mode.
6 Mar 1978	110-73-31	Gary Krier	Evaluation of handling qualities using new tape release intended to improve roll response, and RAV evaluation using ground playback pulse utilizing uplink, formation, gunsight tracking, and ride smoothing.	-First flight to be "pulsed" by pilot with pulses recorded and "played back" to the F-8. -Bank angle control was "significantly improved over all previous flights."
7 Mar 1978	111-74-32	Tom McMurtry	Roll handling qualities evaluation, altitude hold mode evaluation, formation flying characteristics, and gunsight tracking.	-Roll control was further improved. -Lateral axis in formation improved.
8 Mar 1978	112-75-33	Gary Krier	RAV evaluation of 3 pitch filters and pulse playback, autopilot evaluation, formation and tracking.	-Filters allowed quicker response coming out of deadband than basic CAS mode.

24 Mar 1978	113-76-34	Tom McMurtry	Evaluation of transport delay and sample rate with RAV mode, center stick and side-stick for landing approach task with wing down (simulate Shuttle approach).	-Landing gear doors removed to allow higher approach speed, which caused vibration and loss of TM signal to RAV.
25 Mar 1978	114-77-35	Gary Krier	Evaluation of transport delay and sample rate with RAV mode, center stick and side-stick for landing approach task with wing down (simulate Shuttle approach).	-RAV disengage caused 1g jolt.
27 Mar 1978	115-78-36	E. Enevoldson	Pilot familiarization	
28 Mar 1978	116-79-37	John Manke	Pilot familiarization	
7 Apr 1978	117-80-38	Tom McMurtry	Evaluation of effect of transport delay in pitch and roll axis for formation task and approach/landing task.	-Landing task was high-energy wing-down approach.
7 Apr 1978	118-81-39	Gary Krier	Evaluation of effect of transport delay in pitch and roll axis for formation task and approach/landing task.	-Landing task was high-energy wing-down approach. -Delays varied from 20 to 200 msec. -Metronome signals used in frequency response attempts.
10 Apr 1978	119-82-40	Einar Enevoldson	Evaluation of effect of transport delay in pitch and roll axis for formation task and approach/landing task	-Landing task was high-energy wing-down approach
11 Apr 1978	120-83-41	John Manke	Evaluation of effect of transport delay in pitch and roll axis for formation task and approach/landing task.	-Landing task was high-energy wing-down approach -Turbulent air caused alpha light to illuminate, forcing software to configure

				to a fixed set of conditions.
12 Apr 1978	121-84-42	Tom McMurtry	Evaluation of effect of transport delay in pitch and roll axis for formation task and approach/landing task.	-Landing task was high-energy wing down approach -Wing and gear were down for all approaches.
12 Apr 1978	122-85-43	Gary Krier	Evaluation of effect of transport delay in pitch and roll axis for formation task and approach/landing task.	-Landing task was high-energy wing down approach. -Wing and gear were down for all approaches.
14 Apr 1978	123-86-44	Einar Enevoldson	Evaluation of effect of transport delay in pitch and roll axis for formation task and approach/landing task.	-Landing task was high-energy wing down approach. -Delays of 20, 60, and 100 msec.
17 Apr 1978	124-87-45	John Manke	Evaluation of effect of transport delay in pitch and roll axis for formation task and approach/landing task.	-Landing task was high-energy wing down approach. -Wing fuel problems were noted.
18 Apr 1978	125-88-46	John Manke	Evaluation of effect of transport delay in pitch and roll axis for formation task and approach/landing task.	-Landing task was high-energy wing down approach. -No fuel present in wings.
19 May 1978	126-89-47	Tom McMurtry	Evaluating delays for improved CAS and SAS in formation task.	-Flight postponed to this day because of dragging brake pucks.
25 May 1978	127-90-48	Einar Enevoldson	Evaluating delays for SAS and improved CAS in formation task.	
1 Jun 1978	128-91-49	Einar Enevoldson	Landing task using improved CAS or SAS.	-Simulated refueling task also performed.
1 Jun 1978	129-92-50	Gary Krier	Landing task using	-Simulated refueling

				improved CAS or SAS.	task also performed. -CAS appeared to be more responsive than SAS.
14 Jun 1978	130-93-51	Gary Krier		Evaluated effect of lowering control law sample rate utilizing RAV.	-Handling became worse as sample rate decreased.
29 Aug 1978	131-94-52	Gary Krier		Evaluated low sample rate (RAV).	
11 Sep 1978	132-95-53	Tom McMurtry		Evaluated low sample rate (RAV).	-Computer channel "A" failed.
20 Sep 1978	133-96-54	Tom McMurtry		Evaluated low sample rate (RAV).	-Previous abort due to RAV problems.
26 Sep 1978	134-97-55	Einar Enevoldson		Evaluated low sample rate (RAV).	
3 Oct 1978	135-98-56	Tom McMurtry		Evaluated low sample rate (RAV).	
10 Oct 1978	136-99-57	Einar Enevoldson		Evaluate low sample rate (RAV).	
17 Oct 1978	137-100-58	Gary Krier		RAV monitor flt. for RAV adapt.	
24 Oct 1978	138-101-59	Gary Krier		RAV monitor flt. for RAV adapt.	
25 Oct 1978	139-102-60	Gary Krier		RAV adaptive flt. test.	-Computer B failure, generator failure.
3 Nov 1978	140-103-61	Gary Krier		RAV adaptive flt. test.	
8 Nov 1978	141-104-62	Gary Krier		RAV adaptive flt. test.	
14 Nov 1978	142-105-63	Gary Krier		RAV adaptive flt. test.	
15 Nov 1978	143-106-64	Gary Krier		RAV adaptive flt. test.	
20 Nov 1978	144-107-65	Gary Krier		RAV adaptive flt. test.	
21 Nov 1978	145-108-66	Steve Ishmael		Pilot checkout	
21 Feb 1979	146-109-67	Gary Krier		Vibration data on pallet.	
28 Feb 1979	147-110-68	Gary Krier		Vibration data on pallet.	

Date	Number	Pilot	Purpose	Notes
26 Sep 1979	148-111-69	Steve Ishmael	ARM flight evaluation.	-Computer failed in-flight.
31 Oct 1979	149-112-70	Steve Ishmael	ARM flight evaluation.	
21 Nov 1979	150-113-71	Steve Ishmael	ARM flight evaluation.	
27 Nov 1978	151-114-72	Steve Ishmael	ARM flight evaluation.	
29 Nov 1979	152-115-73	Steve Ishmael	ARM flight evaluation.	
7 Feb 1980	153-116-74	Steve Ishmael	ARM flight evaluation.	-Maneuvering flight.
9 Jun 1980	154-117-75	Steve Ishmael	PIO transport and delay evaluation.	
9 Jun 1980	155-118-76	Steve Ishmael	PIO transport and delay evaluation.	
11 Jun 1980	156-119-77	Einar Enevoldson	PIO transport and delay evaluation.	
17 Jun 1980	157-120-78	Tom McMurtry	PIO transport and delay evaluation.	
19 Jun 1980	158-121-79	Steve Ishmael	PIO transport and delay evaluation.	
23 Jun 1980	159-122-80	Steve Ishmael	PIO transport and delay evaluation.	
3 Jul 1980	160-123-81	Steve Ishmael	PIO transport and delay evaluation.	
17 Jul 1980	161-124-82	Steve Ishmael	PIO transport and delay evaluation.	-FCS generator failed at 17,000 ft.
21 Aug 1980	162-125-83	Steve Ishmael	PIO transport and delay evaluation.	
4 Sep 1980	163-126-84	Steve Ishmael	PIO transport and delay evaluation.	
5 Sep 1980	164-127-85	Steve Ishmael	PIO transport and delay evaluation.	
11 Sep 1980	165-128-86	Steve Ishmael	PIO transport and delay evaluation.	

Date	Number	Pilot	Purpose	Notes
24 Sep 1980	166-129-87	Mike Swann	Checkout flt.	
2 Oct 1980	167-130-88	Mike Swann	PIO transport and delay evaluation.	
15 Oct 1980	168-131-89	Mike Swann	PIO transport and delay evaluation.	
24 Oct 1980	169-132-90	Mike Swann	ARM flt. evaluation.	-Channel "A" transient.
29 Oct 1980	170-133-91	Steve Ishmael	ARM flt. evaluation.	-Maneuvering flt., channel "A" transient (2).
31 Oct 1980	171-134-92	Mike Swann	PIO transport and delay evaluation.	
9 Mar 1981	172-135-93	Steve Ishmael	ARM flt. evaluation.	
11 May 1981	173-136-94	Steve Ishmael	PIO transport and delay evaluation.	-New gyro problem.
13 May 1981	174-137-95	Mike Swann	PIOs-gyro data.	
20 May 1981	175-138-96	Steve Ishmael	PIOs (bending platform evaluation).	-Transient fail-"A," replaced "A" platform gyro.
22 May 1981	176-139-97	Mike Swann	PIOs.	-Returned early because of an apparent trim problem.
28 May 1981	177-140-98	Mike Swann	PIOs.	-Aeroman-loop.
11 Jun 1981	178-141-99	Tom McMurtry	ARM.	-Computer "C" fail-S/N08.
24 Jun 1981	179-142-100	Steve Ishmael	ARM.	-Returned early because of TM problem on airplane.
17 Jul 1981	180-143-101	Steve Ishmael	ARM.	
30 Jul 1981	181-144-102	Steve Ishmael	PIOs.	
31 Jul 1981	182-145-103	Steve Ishmael	PIOs.	
4 Aug 1981	183-146-104	Tom McMurtry	PIOs.	-Channel "A" transient.
5 Aug 1981	184-147-105	Steve Ishmael	PIOs.	

Date	Number	Pilot	Purpose	Notes
11 Aug 1981	185-148-106	Tom McMurtry	PIOs; stab. and control.	
14 Aug 1981	186-149-107	Steve Ishmael	PIOs; stab. and control.	
7 Dec 1981	187-150-108	Steve Ishmael	Checkout flt. for A/C systems prior to new pilots flying the A/C.	
8 Dec 1981	188-151-109	Bud Isles	General evaluation flt.	-Grumman forward swept wing pilot.
14 Dec 1981	189-152-110	Major Harry Heimple	General evaluation flt.	-AFTI/F-16 pilot, roll "B" elec; rt. roll "B" transient reset "OK."
8 Jan 1982	190-153-111	Dick Gray	General evaluation flt.	-Computer channel "C" (S/N05) 113 alarm (CPU main storage parity error).
21 Jan 1982	191-154-112	Dick Gray	General evaluation flt.	-Returned early because of "B" bus problem.
2 Feb 1982	192-155-113	Tom McMurtry	CADRE checkout flt.	
9 Feb 1982	193-156-114	Steve Ishmael	CADRE evaluation.	
19 Feb 1982	194-157-115	Steve Ishmael	CADRE evaluation.	-During pre-flight "B" attitude gyro problem, recycle power-OK. Channel "A" transient after landing.
1 Mar 1982	195-158-116	Einar Enevoldson	CADRE evaluation.	
8 Mar 1982	196-159-117	Tom McMurtry	CADRE evaluation.	
9 Mar 1982	197-160-118	Tom McMurtry	CADRE evaluation.	
16 Apr 1982	198-161-119	Dick Gray	Fuel probe check flight.	
6 May 1982	199-162-120	Dick Gray	PIOs-Fuel probe with Navy A-6.	
13 May 1982	200-163-121	Dick Gray	PIOs.	-Navy A-6 abort so returned early

					without accomplishing objective.
19 May 1982	201-164-122	Dick Gray	PIOs (A-6).		-Minor problem with roll "B" elec. right roll servo during preflight. OK during flight.
21 May 1982	202-165-123	Dick Gray	PIOs (A-6).		-Minor problem with roll "B" elec. right roll servo during preflight. OK during flight.
9 Jun 1982	203-166-124	Dick Gray	CADRE 3 evaluation.		
11 Jun 1982	204-167-125	Dick Gray	CADRE 3 evaluation.		
16 Jun 1982	205-168-126	Steve Ishmael	CADRE 3 evaluation.		
17 Aug 1982	206-169-127	Dick Gray	CADRE 3 evaluation.		
30 Aug 1982	207-170-128	Dick Gray	CADRE 3 evaluation.		
4 Jan 1983	208-171-129	Rogers Smith	Familiarization flt.		-Channel "A" transient 2 times in flight.
4 Jan 1983	209-172-130	Rogers Smith	Familiarization flt.		-Channel "C" failure in flight.
20 Jan 1983	210-173-131	Rogers Smith	CADRE 3 evaluation.		-During preflt. found yaw limit problem.
28 Jan 1983	211-174-132	Rogers Smith	CADRE 3 evaluation.		
2 Feb 1983	212-175-133	Rogers Smith	CADRE 3 evaluation.		
10 Feb 1983	213-176-134	Rogers Smith	CADRE 3 evaluation.		-TM dropouts in system 2.
11 Feb 1983	214-177-135	Ed Schneider	Familiarization flt.		
25 Feb 1983	215-178-136	Rogers Smith	Optimal inputs.		
8 Mar 1983	216-179-137	Ed Schneider	Optimal inputs.		
16 Mar 1983	217-180-138	Rogers Smith	CADRE 3, Phase 2.		
17 Mar 1983	218-181-139	Rogers Smith	CADRE 4, Phase 2.		
22 Mar 1983	219-182-140	Rogers Smith	CADRE 4, Phase 2.		

Date	Number	Pilot	Description	Notes
22 Mar 1983	220-183-141	Ed Schneider	CADRE 4, Phase 2.	
12 May 1983	221-184-142	Rogers Smith	CADRE 5, Phase 2.	-Attitude gyro problem, system "A."
13 May 1983	222-185-143	Ed Schneider	CADRE 5, Phase 2.	-Chase (T-38) lost left engine during climb out—returned to base with no data.
18 May 1983	223-186-144	Rogers Smith	CADRE 5, Phase 2.	
20 May 1983	224-187-145	Ed Schneider	CADRE 5, Phase 2.	
24 May 1983	225-188-146	Ed Schneider	CADRE 5, Phase 2.	
26 May 1983	226-189-147	Ed Schneider	CADRE 5, Phase 2.	
24 Jan 1984	227-190-148	Rogers Smith	Functional check flt. and familiarization.	
31 Jan 1984	228-191-149	Ed Schneider	Familiarization flt.	
15 Feb 1984	229-192-150	Ed Schneider	Optimal input assessment.	
9 Mar 1984	230-193-151	Ed Schneider	Optimal input assessment.	
23 Jul 1984	231-194-152	Ed Schneider	REBUS checkout and familiarization flt.	
27 Jul 1984	232-195-153	Rogers Smith	REBUS checkout and familiarization flt.	
22 Aug 1984	233-196-154	Ed Schneider	REBUS flt. 3.	
4 Sep 1984	234-197-155	Rogers Smith	REBUS flt. 4.	
3 Jan 1985	235-198-156	Rogers Smith	Lat. dir. handling, qualities (LDHQ).	
29 Jan 1985	236-199-157	Ed Schneider	LDHQ.	
1 Feb 1985	237-200-158	Ed Schneider	REBUS Flt. 5.	
8 Mar 1985	238-201-159	Rogers Smith	OPTRE.	
11 Mar 1985	239-202-160	Rogers Smith	OPTRE.	
11 Mar 1985	240-203-161	Rogers Smith	OPTRE.	-No wing fuel.

12 Mar 1985	241-204-162	Ed Schneider	OPTRE.	-No wing fuel.
12 Mar 1985	242-205-163	Ed Schneider	OPTRE.	-No wing fuel.
13 Mar 1985	243-206-164	Rogers Smith	OPTRE.	-No wing fuel.
13 Mar 1985	244-207-165	Rogers Smith	OPTRE.	-No wing fuel.
19 Mar 1985	245-208-166	Rogers Smith	OPTRE.	-No wing fuel.
26 Mar 1985	246-209-167	Rogers Smith	OPTRE.	
2 Apr 1985	247-210-168	Ed Schneider	REBUS flt. 6.	-No wing fuel. -After-burner problems noted.
16 Dec 1985	248-211-169	Ed Schneider	Functional check flight.	-Aileron feel system modification for this flight.

Glossary

Actuator—a mechanical device that moves some other component on command.

Ailerons—surfaces, usually on the trailing edge of the wings, that control rolling an airplane.

Airfoil—a wing or other surface that provides lift.

Amplitude—the vertical range of an oscillation.

Analog—something that models something else; in analog flight-control computers, an electronic circuit that solves equations of motion that model aircraft maneuvers.

Area-rule—a way of shaping an airplane's fuselage that reduces drag.

Analytic Redundancy Management (ARM)—a method of redundancy management by analyzing and fusing data collected from different non-redundant sensors.

Attitude controller—a short control stick for commanding a change in the position of a spacecraft relative to its direction of flight.

Automatic interception interface—the interface between a ground control radar station and a fighter interceptor.

Avionics—electronics used in aircraft.

Backup Control System (BCS)—the analog flight computers and their associated actuators, in the case of the F-8 Digital Fly-By-Wire aircraft's Phase I configuration.

Ballistic—the essentially parabolic path taken by an object when it has been accelerated and then the acceleration ceases.

Baseline loop—the length of time devoted to a cycle in a control system.

Boolean function—a mathematical function that returns a reading of either "true" or "false."

Breadboard—a prototype of a computer or other electronic device built by the design group to test the device before it is packaged for production.

British "Oboe"—a method of guiding bombers to their targets via a directional radio beam.

Canards—horizontal lifting and control surfaces placed in front of the wings.

Centerline air scoop—an air intake for a turbojet engine mounted in the center of a fuselage.

Command Augmentation System (CAS)—an automated system for controlling the flight-control surfaces, used only in the pitch axis on the F-8. It could predict and smooth out pitch oscillations.

Compiler—a computer program that accepts statements of a high-level language as input and generates machine code that will execute those statements as output.

Computer Bypass System (CBS)—the backup analog computer-based flight-control system used in Phase II of the F-8 project.

Control laws—aircraft equations of motion encapsulated in either analog circuits or software.

Cruciform tail—the stabilizing surfaces of an airplane arranged in the shape of a cross. This arrangement of horizontal and vertical stabilizers is the most common among all aircraft.

Damper—something that reduces oscillations.

Delta wing—the triangular shaped wing used on many jet aircraft, often without any horizontal tail surface as a stabilizer.

DIR—the direct control mode of the fly-by-wire system.

Discrete-circuit transistorized computer—a computer constructed of individual transistors and other electronic components instead of integrated circuits.

Discretes—values that signal an event or a state.

Downlink—the radio connection used to send information from an aircraft or spacecraft to the ground.

DSKY (Display and Keyboard Unit)—a component of the Apollo spacecraft that enabled input and output to the computer system. One was used on the F-8 during Phase I of the fly-by-wire research program.

Drag—the resistance on an aircraft caused by moving through the air.

Dynamic stability—stability maintained by a flight-control system.

Effectors—devices that act on other devices or perform some service.

Egon—see German "Egon."

Electrical analog device—a circuit that models behavior.

Elevators—control surfaces that move an aircraft in the pitch axis.

Fault tolerance—the ability of a device to function after a failure.

Fixed gain stability augmentation system—an augmentation system that uses a single set value to condition inputs.

Fixed-point arithmetic—the representation of numbers using an immovable decimal point.

Flight pallet—the platform on which the computers and other devices could be mounted for installation in the airplane.

Floating-point arithmetic—the representation of numbers using a movable decimal point.

Fuselage—the body of an aircraft.

g—a force equal to that of the gravity of the Earth at sea level.

g tolerance—the ability to withstand force on the body or on a structure caused by aircraft maneuvering.

Gain—a predefined coefficient that is applied in the control laws of a fly-by-wire aircraft to affect the sensitivity of the results of a command. The values of the gains could be altered in software and a range of gains could be selected using rotary switches on the mode control panel.

Gate—a logic device. For example, an AND gate returns the result of a Boolean AND operation on its inputs.

German "Egon"—a method of guiding bombers to their targets via a directional radio beam.

Gimbal—a device with two axes of rotation in which a gyroscope is mounted.

GCA (Ground controlled approach)—a method of guiding an aircraft to the runway in bad weather using radar and a controller who radios instructions to the pilot.

Gyroscope—a rotating device that provides a reference for instruments and imparts stability.

Hardover—the rapid deflection of a control surface to its physical stopping point.

High gain model following—a method of monitoring performance by comparing actual values to optimal predicted values.

High-g cockpits—cockpits designed to help pilots withstand g forces.

Horizontal stabilizer—an airfoil, usually mounted toward the tail, that often provides downward forces to help balance an aircraft. On a fly-by-wire aircraft, it can be a lifting surface instead.

Inner loop—a real-time control program sometimes has fixed time periods for the execution of blocks of software. These periods are usually of differing lengths, and the shortest is the inner loop.

Instability—the lack of ability to remain stable.

Integrated circuit—an electronic circuit containing many transistors and other components installed on a small silicon wafer.

Integrated engine/flight controls—controls that can affect both the engine and control surfaces together, such as throttle control and flap deployment on an approach.

Integrator—an electronic circuit that performs the integration operation of the calculus.

Interceptor—a fighter designed to find and destroy enemy aircraft.

Interface—the connection between two devices for the exchange of data.

KECO—the cooling system used in the Phase I flight pallet to keep the electronic devices within temperature limits.

KIAS (knots indicated air speed)—the aircraft speed shown on an instrument in the cockpit, uncorrected for the effects of wind and often expressed in terms of nautical miles per hour.

K-START—the name of the contents of the magnetic tape used during Phase I that contained the constants and additional software for a specific flight.

Lateral stability—the ability to maintain wings-level flight.

Learning algorithm subsystem—a control program that automatically changes its output based on information gained during its operation.

Lift—the upward force provided by airfoils moving through the air.

Linear variable differential transformer (LVDT)—a device that converts physical force into a proportional voltage.

Logic circuits—electronic circuits that represent some Boolean equation.

Longitudinal stability—the ability to remain stable in the pitch axis.

Lyapunov model reference technique—a mathematical technique used to monitor performance by comparing actual values to optimal predicted values.

Mach—the aircraft velocity relative to the speed of sound.

Majority logic voters—circuits that return a value based on examining inputs and choosing the one that represents the majority.

Mockups—non-functional models, often actual size, that make it possible to check component positions, sizes, wiring lengths, etc. before committing to a final design.

Model tracking—a method of prediction arrived at by monitoring how well something follows a model of its motion, then projecting future activity.

Monoplanes—airplanes with one set of wings.

Newton-Raphson technique—a mathematical technique used to monitor performance by comparing actual values to optimal predicted values.

NOR operation—a Boolean operation that negates the result of an OR operation. Thus, if the inputs to a memory chip were one, one, and zero, the output of the integrated circuit would be zero.

OPTRE (Optimum Trajectory Research Experiment)—a Phase II flight experiment that involved testing data uplink and downlink between the F-8 and a computer in the new Remotely Piloted Vehicle Facility.

OR—a Boolean operation in which if any of the inputs are ones, a one is the result.

Orthogonal—in a perpendicular direction.

Outer loop—a real-time control program sometimes has fixed time periods for the execution of blocks of software. These periods are usually of differing lengths, and the longest is the outer loop.

Pilot-induced oscillation (PIO)—what happens when a pilot tries to do a maneuver but uses too much force on the controls and the aircraft overshoots the desired attitude; then if the pilot tries to recover from this, but overcorrects, thus forcing even more recovery cycles, the condition is called a PIO. The current terminology is "airplane-pilot coupling."

Piston-driven airplane—an airplane powered by a reciprocating engine similar to that used by an automobile.

Pitch axis—the axis about which the nose of an airplane appears to move up and down.

Pitch evaluation—maneuvers to test a control system's stability in the longitudinal axis.

REBUS (Resident Backup Software)—a means of providing software safety by including a kernel capable of controlling the aircraft as a backup within the full load of software.

Recursive least-squares approach—a mathematical technique used to obtain optimal predicted values.

Redundancy—the duplication of components so that a failed one can be ignored and the flight continued using the duplicates.

Remotely Augmented Vehicle (RAV) mode—a procedure in which the control laws are executed on the ground and the commands sent up to the aircraft.

Roll axis—the axis around which an airplane appears to rotate.

Roll steps—maneuvers in which test pilots make rolls but stop at intervals instead of smoothly rotating through the entire axis.

RRC (Roll Rate Control)—a mode only selectable while using the Command Augmentation System, which only worked in the pitch axis; RRC nevertheless gave additional control in the roll axis.

Rudder—a control surface that helps move the airplane in the yaw axis.

Rudder pulses—maneuver by which test pilots make short pushes on the rudder controls to check stability in the yaw axis.

Sample rates—the number of separate values returned by a sensor in a fixed time.

Sensor analytic redundancy management—schemes for figuring out which one of the sensors returning different values is correct.

Sensors—devices that measure things like airspeed, attitude, and acceleration and return values to the control system.

Sensor suite—the set of sensors on a particular aircraft.

Servo—a device that executes commands from the control system and moves other devices.

Side-stick controller—a control device mounted on the side of the cockpit rather than in the center.

Simplex-with-backup—a single control string with a dissimilar system as an alternate in case of failure.

Single-string—a system consisting of only one of every needed component.

Slider valves—devices in an hydraulic system that move in response to pressure changes.

Sperry flight-control equipment—analog computers and other control devices developed by the Sperry company.

Stability augmentation system (SAS)—a mode that provides automatic help to a pilot to maintain control in gusts and to reduce the probability of pilot induced oscillations.

Stealth—the ability of an aircraft to avoid radar detection.

Swept-wing—wings mounted at other than a 90-degree angle to the fuselage.

Telemetry—signals sent from an aircraft or spacecraft to the earth containing data gathered or generated by experiments and flight hardware.

Three-channel redundant analog computer—a computer system using analog circuits in which all are triplicated for redundancy.

Thrust—force generated by an engine.

Titan booster—a rocket used to launch piloted and unpiloted earth satellites.

Trim control—a device that enables small control surface deflections to maintain an aircraft in a desired attitude by compensating for changes in the position of the center of gravity and winds.

Uplink—the transmission of signals from the ground to a vehicle in flight.

Variable gains—different selectable values for conditioning inputs.

Variable stability—stability that changes from one type to another as effected by the control system.

Ventral—mounted downward.

Vertical stabilizer—a stabilizing surface mounted mostly perpendicular to a horizontal stabilizer or wing.

Voters—devices that return the value of the majority of a set of inputs.

Wing root—the mounting point of a wing.

Wind tunnel—a device that accelerates air past a model of an aircraft (or in some cases, an actual aircraft) for research and development purposes.

Yaw axis—the axis about which the nose of an airplane appears to move side to side.

Bibliography

A Note on Sources

This book is based on the plethora of sources generated by any NASA technical project or available after its completion: personal interviews and a wide variety of papers, books, and manuscripts. In this bibliography is a list of the interviews that serve as primary sources and published or manuscript sources. Unpublished memos, flight reports, personal logs, etc. are in one of two locations. The Dryden Flight Research Center History Office has the majority and the James E. Tomayko Collection of NASA Documents (MS87-8) in the Special Collections Department of the Ablah Library at The Wichita State University has Apollo and some Shuttle materials. Unless otherwise indicated, the author conducted all of the interviews.

Interviews

Burke, Melvin, Palmdale, CA, 26 Mar. 1998; telephone interview, 17 Feb. 1998.

Deets, Dwain A., Dryden Flight Research Center, 5 Jan. 1998.

Felleman, Philip, Draper Laboratory, 27 May 1998.

Hirschler, Otto, telephone interview, Huntsville, AL, Oct. 1983.

Hughes, Frank, Johnson Space Center, Houston, TX, 2 June 1983.

Jarvis, Calvin, Lancaster, CA, 7 January 1998; telephone interviews, 1 June 1998, 19 September 1998; with Lane Wallace, Dryden Flight Research Center, 30 Aug. 1995.

Krier, Gary, Dryden Flight Research Center, 9 January 1998; telephone interview, 24 July 1998.

Lock, Wilton P., Dryden Flight Research Center, 25 March 1998; telephone interview with John "Dill" Hunley, Dryden Flight Research Center, 14 Oct. 1998.

McRuer, Duane, interview with Lane Wallace, Hawthorne, CA, 31 Aug. 1995.

Megna, Vincent, Draper Laboratory, 27 May 1998.

Morris, James, Dayton, Ohio, May 1990; telephone interview, Oct. 1995.

Phelps, James R., Dryden Flight Research Center, 27 Mar. 1998.

Szalai, Kenneth R., Dryden Flight Research Center, 8 June 1998, 12 June 1998; telephone interview 30 Sept. 1998; with Lane Wallace, Dryden Flight Research Center, 30 Aug. 1995.

Tarnowsky, Etienne, Airbus, telephone interview, Toulouse, France, 4 Feb. 1998.

Wilson, Ronald J. "Joe," Dryden Research Center, 10 June 1998.

Printed or Manuscript Sources

Anderson, D.C., and R.L. Berger. "Maneuver Load Control and Relaxed Static Stability Applied to a Contemporary Fighter Aircraft." AIAA Paper 72-87, 1972.

Aplin, John. "Primary Flight Computers for the Boeing 777." Manuscript.

Armstrong, Neil, personal communication, 8 Mar. 1998.

Bairnsfather, Robert R. "Man-Rated Flight Software for the F-8 DFBW Program," in *Description and Flight Test Results of the NASA F-8 Digital Fly-By-Wire Control System*. Washington, DC: NASA TN D-7843, 9-11 July 1974, pp. 93-126.

Barnwell, F.S. *Airplane Design*. New York: Robert M. McBride and Co., 1917.

Beniger, James R. *The Control Revolution*. Cambridge, MA: Harvard University Press, 1986.

Bernhard, Robert. "All Digital Jets on the Horizon," *IEEE Spectrum* (Oct. 1980): 38-41.

Boeing. "Sneak Circuit Analysis of F-8 Digital Fly-by-Wire Aircraft," D2-118582-1, Mar. 1976.

Breuhaus, Waldeman O. "The Variable Stability Airplane from a Historical Perspective." Unpublished manuscript, 26 Feb. 1990.

Bright, James R. "The Development of Automation," pp. 635-655 of Melvin Kranzberg and Carroll W. Pursell, Jr., eds. *Technology in Western Civilization*. New York: Oxford University Press, 1967.

Brown, Samuel R., R.R. Larson and Kenneth J. Szalai. "F-8 Digital Fly-By-Wire Active Control Law Development and Flight Test Results." Manuscript, 26 Aug. 1977.

Bryan, G.H., and W.E. Williams. "The Longitudinal Stability of Aerial Gliders," in *Proceedings of the Royal Society of London*, 73 (1904): 100-116.

Condom, Pierre. "Systems for the Airbus A320 (Innovation in all Directions)," *INTERAVIA*, 4 (1985): 353-355.

Daley, Ed, et al. "Unstable Jaguar Proves Active Controls for EFA,"*Aerospace America*, May 1985.

Deckert, James C. "Analytic Redundancy Management Mechanization and Flight Data Analysis for the F-8 Digital Fly-By-Wire Aircraft Flight Control Sensors." Charles Stark Draper Laboratory, Inc., CSDL-R-1520, Sept. 1982.

Deckert, James C. "Definition of the F-8 DFBW Aircraft Control Sensor Analytic Redundancy Management Algorithm," R-1178, Charles Stark Draper Laboratory, Inc., Aug. 1978.

Deets, D. A. "Design and Development Experience with a Digital Fly-By-Wire Control System in an F-8C Airplane," in *Description and Flight Test Results of the NASA F-8 Digital Fly-By-Wire Control System*. Washington, DC: NASA TN D-7843, 9-11 July 1974, pp. 13-40.

Deets, D. A,. and K. J. Szalai. "Design and Flight Experience with a Digital Fly-By-Wire Control System Using Apollo Guidance System Hardware on an F-8 Aircraft." *AIAA Guidance and Control Conference* (AIAA Paper No. 72-881), 1972.

Deets, Dwain A., and Kenneth J. Szalai. "Phase I F8 DFBW Software Specification." Final Revision, 30 Mar. 1973.

Deyst, J., J. Deckert, M. Desai, and A. Willsky. "Development and Testing of Advanced Redundancy Management Methods for the F-8 DFBW Aircraft." AIAA manuscript, pp. 309-315.

Draper, Charles Stark. "Flight Control." 43rd Wilbur Wright Memorial Lecture, *Journal of the Royal Aeronautical Society*, 59 (July 1955): 451-478.

Dryden, Hugh L., "General Background of the X-15 Research Airplane Project." *Research-Airplane-Committee Report on Conference on the Progress of the X-15 Project*. Langley Aeronautical Laboratory, 25-26 Oct. 1956.

Ettinger, R. C. "The Implications of Current Flight Control Research and Development," in *Society of Experimental Test Pilots*, 15, no. 2 (24-27 Sept. 1980): 18-37.

Felleman, Philip G. "An Aircraft Digital Fly-by-Wire System." Manuscript, delivered at the 29th Annual ION Meeting, St. Louis, MO., June 1973, pp. 1-7.

Fisher, F.A. "Lightning Considerations on the NASA F-8, Phase II Digital Fly-By-Wire System." GE, Pittsfield, MA, June 1975.

Fleck, J. J. and D. M. Merz. "Research and Feasibility Study to Achieve Reliability in Automatic Flight Control Systems." TR-61-264, General Electric Company, Mar. 1961.

Floyd, J. C. "The Canadian Approach to All-Weather Interceptor Development." *Journal of the Royal Aeronautical Society*, 62, no. 576 (Dec. 1958): 845-866.

Gee, Shu W., and Melvin E. Burke. "NASA Flight Research Center Fly-By-Wire

Flight-Test [sic] Program." Briefing slides and commentary, 1971.

General Electric. "Single Failure Point Study of the CSM Guidance and Controls System for the Moon Mission." Contract NASW-410, Apr. 1968.

Gibbs-Smith, Charles H. *Aviation: An Historical Survey From Its Origins to the End of World War II.* London: Her Majesty's Stationery Office, 1970.

Gough, Melvin. "Notes on Stability from the Pilot's Standpoint." *Journal of the Aeronautical Sciences*, 6, no. 10 (Aug. 1939): 395-398.

Hall, Eldon C. "Reliability History of the Apollo Guidance Computer." Draper Laboratory Report R-713, Jan. 1972.

Hartmann, Gary L., and Gunter Stein. "F-8C Adaptive Control Law Refinement and Software Development." Honeywell Systems and Research Center, Minneapolis, MN, June 1977.

Honeywell. "Honeywell Digital Adaptive Control Laws for the F-8." Volume 1, Technical, Minneapolis, MN, 15 Jan. 1974.

Hooker, David S., Robert L. Kisslinger, and George R. Smith. *Survivable Flight Control System Final Report.* Dayton, OH: Air Force Flight Dynamics Laboratory, Air Force Systems Command, 1973.

Hunt, G. H. "The Evolution of Fly-By-Wire Control Techniques in the UK," in *11th International Council of the Aeronautical Sciences Congress*, 10-16 Sept. 1978, pp. 61-71.

Hydraulic Research. Final Design Review Phase II Servoactuators, May 1975.

IBM. AP-101 Technical Description, 7 Oct. 1974. Owego, NY.

Jarvis, Calvin R. "An Overview of NASA's Digital Fly-By-Wire Technology Development Program," in *Description and Flight Test Results of the NASA F-8 Digital Fly-By-Wire Control System*. Washington, DC: NASA Technical Note TN D-7843, 1975.

Jenny, R.B., F.M. Krachmalnik, and S.A. Lafavor. "Air Superiority with Control Configured Fighters." *AIAA Journal of Aircraft* (May 1972).

Keese, W. M., *et al.* "Management Procedures in Computer Programming for Apollo— Interim Report." Bellcomm, Inc., TR-64-222-1, 30 Nov. 1964.

Kisslinger, Robert L., and Robert C. Lorenzetti. "The Fly-By-Wire Systems Approach to Aircraft Flying Qualities," in *NAECON* (15-17 May 1972): 205-210.

Koppen, Otto. "Airplane Stability and Control from a Designer's Point of View,"

Journal of the Aeronautical Sciences,7, no. 4 (Feb. 1940): 135-140.

Kuo, Benjamin C. *Analysis and Synthesis of Sampled-Data Control Systems.* Englewood Cliffs, NJ: Prentice-Hall, 1963.

Kurten, P.M. *Apollo Experience Report: Guidance and Control Systems—Lunar Module Abort Guidance System.* Washington, DC: NASA TN-D-7990, 1975.

Larson, Richard R., Rogers E. Smith, and Keith Krambeer. "Flight Test Results Using Nonlinear Control with the F-8C Digital Fly-By-Wire Aircraft." AIAA Paper 83-2174, Aug. 1983.

Lawton, T. J., and C. A. Muntz. "Apollo Guidance and Navigation." E-1758, MIT, May 1965.

Lock, Wilton P., William R. Peterson, and Gaylon B. Whitman. "Mechanization of and Experience with a Triplex Fly-By-Wire Backup Control System," in *Description and Flight Test Results of the NASA F-8 Digital Fly-By-Wire Control System.* Washington, DC: NASA TN D-7843, 1975: 41-72.

Maxim, Hiram S. *Artificial and Natural Flight.* London: Whittaker, 1909.

McRuer, Duane, and Dunstan Graham. "Eighty Years of Flight Control: Triumphs and Pitfalls of the Systems Approach," in *Journal of Guidance and Control*, 4 (July-Aug. 1981): 353-362.

Megna, Vincent A., and Wilton R. Lock. "F-8 DFBW System Redundancy Management Analysis and Resident Backup Software." Manuscript, no date, Dryden Flight Research Center Historical Reference Collection.

Mekel, R., R.C. Montgomery, and S. Nachmias. "Learning Control System Studies for the F-8 DFBW Aircraft." Manuscript, Dryden Flight Research Center Historical Reference Collection.

Milliken, William F. "Progress in Dynamic Stability and Control Research." *Journal of the Aeronautical Sciences*, 14, no. 9 (Sept. 1947): 493-519.

Morisset, J. "Fly-By-Wire Controls are on the Way," *Telonde* (Dec. 1983): 7-9.

Mutchler, J. V., and H. A. Sexton. "APOLLO CMC/LGC Software Development Plan." TRW Note No. 68-FMT-643, Apr. 1968.

NASA. *Description and Flight Test Results of the NASA F-8 Digital Fly-By-Wire Control System.* Washington DC: NASA TN D-7843, Feb. 1975.

NASA Fact Sheet FS-802-84-1. *"Fly By Wire."* 24 Jan. 1984.

NASA. "Government/Industry Workshop on Methods for the Certification of Digital

Flight Controls and Avionics." Washington, DC: NASA TM X-73,174. Oct. 1976.

NASA. *Proceedings of the F-8 Digital Fly-By-Wire and Supercritical Wing First Flight's 20th Anniversary Celebration.* Washington, DC: NASA Conference Publication 3256, 27 May 1992. Volumes I & II.

Naur, Peter, and Brian Randell. *Software Engineering, a Report on a Conference Sponsored by the NATO Science Committee.* Garmisch, Germany, 7-11 Oct. 1968, (Jan. 1969).

Naval Air Systems Command. *NATOPS Flight Manual, Navy Model F-8C and F-8K Aircraft.* NAVAIR 01-45HHC-501, 1 Jan. 1970, revised 15 Jan. 1971.

Paulk, M., et al. Capability Maturity Model for Software. Software Engineering Institute, Carnegie Mellon University, Pittsburgh, PA, 1993.

Petersen, Kevin. "F-8 RAV Baseline Experiment." Preliminary Report S75-15-014, 31 Oct. 1975.

Petersen, Kevin, and Kenneth J. Szalai. "F-8 Software Release Documentation," F8-77-014, 3 Aug. 1977.

Plumer, J. Anderson, Wilbert A. Malloy, and James B. Craft. "The Effects of Lightning on Digital Flight Control Systems," in *Description and Flight Test Results of the NASA F-8 Digital Fly-By-Wire Control System.* Washington, DC: NASA TN D-7843, Feb. 1975.

Powers, B.A., and Kenneth J. Szalai. "Adaptive Control Flight Test Completion." Milestone Report F8-78-029, 17 Dec. 1978.

Preliminary Pilot's Operating Instructions Arrow 1. Malton, Ontario: Avro Aircraft Limited, Apr. 1958.

Reed, Fred. "The Electric Jet." *Air and Space*, 1 (Dec. 1986 - Jan. 1987): 42-48.

Reed, R. Dale, with Darlene Lister. *Wingless Flight: The Lifting Body Story.* Washington, DC: NASA SP-4220, 1997.

Reilly, T.J., and J.S. Prince "Relative Merits of Digital and Analog Computation for Fly-By-Wire Flight Control," in J.P. Sutherland. *Proceedings of the Fly-By-Wire Flight Control System Conference.* Wright Patterson Air Force Base, OH: AFFDL-TR-69-58, 1969: 203-225.

"Resident Backup Software (REBUS) Program," manuscript, no date, Dryden Historical Reference Collection.

Roskam, Jay. Lecturer's notes from "Airplane Stability and Control: Past, Present, and Future." Long Island section of *AIAA*, 16 Mar. 1989.

Rushby, John. *Formal Methods and the Certification of Critical Systems.* Menlo Park, CA: SRI International, 1993.

Saltzman, Edwin J., and Theodore G. Ayers. *Selected Examples of NACA/NASA Supersonic Flight Research.* Edwards, CA: NASA SP-513, 1995.

Sayers, W.H. "A Simple Explanation of Inherent Stability," in F.S. Barnwell. *Airplane Design.* New York: Robert M. McBride and Co., 1917, pp. 73-102.

Schad, Leslie A., "DO-178B, Software Considerations in Airborne Systems and Equipment Certification." Manuscript, 29 Dec. 1997.

Sperry Flight Systems. "Description and Theory of Operation of the Computer Bypass System for the NASA F-8 Digital Fly-By-Wire Control System." Phoenix, AZ: Sperry Flight Systems, Dec. 1976.

Susskind, Charles. *Understanding Technology.* Baltimore, MD: John Hopkins University Press, 1973.

Sutherland, J. P. "Introduction to Fly-by-Wire." *Proceedings of the Fly-By-Wire Flight Control System Conference.* Dayton, OH: Air Force Flight Dynamics Laboratory, Air Force Systems Command, 1969, p. 14.

Sweetman, Bill, and James Goodall. *Lockheed F-117A: Operation and Development of the Stealth Fighter.* Osceola, WI: Motorbasics, International, 1990.

Szalai, Kenneth J. "F-8 Digital Fly-By-Wire Systems/Software Test Experience." Manuscript in Dryden Historical Reference Collection, Sept. 1976.

Szalai, Kenneth J., Philip G. Felleman, Joseph Gera, and R. D. Glover. "Design and Test Experience with a Triply Redundant Digital Fly-By-Wire control System." AIAA Paper 76-1911. *Proceedings, Guidance and Control Conference,* San Diego, CA, Aug. 16-18, 1976.

Szalai, Kenneth J., and Joseph Gera. F-8 Digital Fly-By-Wire Phase II Software Specification. Revision F, NASA Dryden Flight Research Center, Edwards, California, 1 Dec. 1975.

Szalai, Kenneth J., and Vincent A. Megna. *R-1164 Digital Fly-By-Wire Flight Control Validation Experience.* Cambridge, Massachusetts: The Charles Stark Draper Laboratory, Inc., 1978.

Taylor, J. C. "Fly-By-Wire and Redundancy," in J.P. Sutherland ed., *Proceedings of the Fly-By-Wire Flight Control System Conference.* Wright-Patterson Air Force Base: Air Force Flight Dynamics Laboratory Technical Report AFFDL-TR-69-58, 16-17 Dec. 1968: 187-202.

Tillman, Barrett. *MiG Master (The Story of the F-8 Crusader)*. Annapolis, MD: Naval Institute Press, 1990.

Tomayko, James E. "Blind Faith: The United States Air Force and the Development of Fly-By-Wire Technology," in *Technology and the Air Force*. Washington D.C.: The United States Air Force, 1997, pp. 163-186.

Tomayko, James E. "Helmut Hoelzer's Fully Electronic Analog Computer." *Annals of the History of Computing*, 7, no. 3 (July 1985): 227-240.

Tomayko, James E. *Computers in Spaceflight*. Washington, DC: NASA Contractor Report-182505, Mar. 1988.

Vetsch, G. J., R. J. Landy, and D. B. Schaefer. "Digital Multimode Fly-By-Wire Flight Control System Design and Simulation Evaluation." AIAA Digital Avionics Systems Conference, 2-4 Nov. 1977, pp. 200-211.

Vincenti, Walter A. *What Engineers Know and How They Know It*. Baltimore, MD: Johns Hopkins University Press, 1990.

von Neumann, John. "Probabilistic Logics and the Synthesis of Reliable Organisms from Unreliable Components. " In William Aspray and Arthur Burks. *Papers of John von Neumann on Computers and Computer Theory*. Cambridge, MA: Charles Babbage Institute Reprint Series for the History of Computing, Vol. 12, The MIT Press, 1987, pp. 553-576.

Wallace, Lane E. *Flights of Discovery: Fifty Years at the NASA Dryden Flight Research Center*. Washington, DC: NASA SP-4309, 1996.

Watson, John H. "Fly-By-Wire Flight Control System Design Considerations for the F-16 Fighter Aircraft." AIAA Paper 76-1915. *Proceedings, Guidance and Control Conference*, San Diego, CA, 16-18 Aug. 1976.

White, Rolland J. "Investigation of Lateral Dynamic Stability in the XB-47 Airplane." *Journal of the Aeronautical Sciences,* 17, no. 3 (Mar. 1950): 133-148.

Wilson, R. E., et al. "APOLLO Guidance Software—Development and Verification Plan," NASA Manned Spacecraft Center, Oct. 1967.

Wimpress, John K., and Conrad F. Newberry. *The YC-14 STOL Prototype*. Reston, VA: American Institute of Aeronautics and Astronautics, 1998.

Wright, Orville. *How We Invented the Airplane*. Edited by Fred C. Kelly. New York: David MacKay, 1953.

About the Author

James E. Tomayko is the Director of the Master of Software Engineering Program at Carnegie Mellon University. He has over 20 years of experience in the computing industry and academia. Besides his technical expertise in software engineering, he holds a doctorate in the history of technology from Carnegie Mellon University, earned in 1980. Before he returned to his alma mater in 1989 to teach computer science, he founded the software engineering graduate program at Wichita State University. He has also been employed by or held contracts with such firms as NCR, Boeing, Ansys, Carnegie Works, Keithley Instruments, ADP, and Mycro-Tek. In addition, he brings to the writing of this book a private pilot certificate with an instrument rating. He is a department editor of the IEEE *Annals of the History of Computing* and has published extensively in that field, including articles in *Aerospace Historian*, *The Journal of the British Interplanetary Society*, and *American Heritage of Invention and Technology*. He is also the author of *Computers in Space: Journeys with NASA* (Indianapolis, IN: Alpha Books, 1994), among other publications.

Index

680J – 65; project, 88; tail number, 62; F-4, 98
A4 Rocket (V-2), 13, 14 ill., 15, 16, 25, 77
Abort Guidance System (AGS), 45
AC Spark Plug, see Delco
Active control, 9-11, 16, 22, 114, 125
Active flight control, 126
Actuator, 27-28, 32 ill., 52, 66-67, 67 ill.
Adaptive Control Laws, 114-118
Adaptive Control Law Set, 115
Advanced Fighter Technology Intergration, see AFTI F-16
Advisory Circular 20-115B, 129
AFTI F-16 (Advanced Fighter Technology Integration), 111, 132 ill.
Air Force, U.S., 1, 17-18, 26, 29-30, 33, 38, 46, 62, 132
Airbus, 122, 125 ill., 128, 133
Airplanes, see individual designations (such as X-24A, F-8)
Aldrin, Edwin "Buzz" E., Jr., 19
AMD 29050 computer, 128
American Airlines, 107
Ames Research Center, 74, 129
Analog, backup system, 61, 80, 97; circuits, 16, 26; computers, 15, 27-30, 37, 52, 62, 64, 100; flight control system, 22-23, 25, 33, 45, 62, 69; instruments, 79; signals, 19; simulator, 49; to-digital, 26
Anomaly, 53
AP-101 (IBM) computer, 91, 92 ill., 93-95, 100, 105, 108-10, 116
Apollo Program, 21, 27, 31, 40, 45, 47; Apollo 9, 18; Apollo 13, 112; Apollo 14, 54; Apollo 15, 61; computer, 31, 33, 35, 46, 48, 55 ill., 61, 64, 67, 70, 90, 115; control systems, 39, 89; DSKY (Display and Keyboard Unit), 39, 40 ill., 47, 59, 61, 71-3; crew of, 41; digital computer system, 59; guidance and navigation systems, 44, 47, 50, 60 ill., 131; inertial measurement unit, 65; inertial navigation system, 52; Lunar Module, 4, 16; project, 134; software, 42-43, 53-54; space program, 61, 87; system, 93
Approach power compensator, 62
Armstrong, Neil, 19, 31, 46
Assembly Control Board, 53
Athans, Dr. M., 116
Atlas booster, 17-18
Automation, 130

Autonetics D-216 computer, 90
Avro, see CF-105

B-1, 10
B-1A, 90
B-2, 122, 127 ill.
B-47, 29 ill., 30
Backup Control System (BCS), 55, 63-4, 70, 73, 76, 78, 80-1
Backup Flight System, 61, 63 ill.
Bairnsfather, Robert "Barney," 53
Barnwell, F.S., 5-6
Bell Aircraft, 21
Bellcomm, Inc., 40, 44
Bicycle, 12 ill.
Bikle, Paul, 28, 30
Bird, Dr. John, 91, 116
Boeing, 7, 17, 24, 100-01, 122, 127-28; Boeing 737, 57, 88; Boeing 747, 107; Boeing 757, 127; Boeing 767, 127; Boeing 777, 35 ill., 128-29, 133; Boeing B-47, 10, 30, 61, 65
Borek, Robert, 109
Brown, Sam R., 97
Bryan, G.H., 3, 5
Building 4800 (at the Flight Research Center), 104
Burke, Melvin E., 22, 26, 28, 30-3, 46, 131
Burser, Q.W. "Jerry," 66

C-141, 65
C-54, 11
California Institute of Technology, 37
Canadian Air Force, 15
Canard, 9 ill.
Capability Maturity Model (CMM), 43
Carnegie Mellon University, 43; Software Engineering Institute, 43
Cayley, George, 2
CF-105, 15-16, 16 ill., 26, 30 n., 45, 63, 65
Chanute, Octave, 3
Cherry, Dr. George, 31
Christensen, John, 94
Collar, A.R., 12
Command Module, 18-9
Computer Bypass System (CBS), 97, 98 ill., 103-4
Computer Interface Panel, 96-7, 107, 109, 113
Concorde, 11, 23

Control augmentation, 11
Control Augmentation System (CAS), 61-2, 72, 75-6, 81, 96, 114
Control configured vehicle, 91, 109
Control Data Corporation Alpha, 89
Control law, 25, 29, 48, 50-2, 54, 97, 103, 108-9, 112; active, 115; adaptive, 85, 114, 116, 118; advanced, 100; analytic, 85; development, 49
Convair CV-990, 74
Cooper-Harper Scale, 70, 75, 81, 101
Cooperative Advanced Digital Research Experiment (CADRE), 111, 114
Cornell Aeronautical Laboratory, 22, 50
Cox, Dr. Kenneth, 85, 91
Craft, James B., 71, 73, 78, 104, 109
Critical design review, 42
Curtis JN-5 Jenny, 5
Customer Acceptance Readiness Review (CARR), 42-43
CYBER computer, 116
Cybernetics, 130

da Vinci, Leonardo, 1
Dana, William H., 82
DC-9, 87-8
Deadband, 50-1
Debus, Kurt, 13
Deckert, James C., 118
Deets, Dwain A., 22, 30, 49-50, 52, 73, 116, 120, 134
Delco (formerly AC Spark Plug), 48, 59, 66-7
Differential analyzer, 25-6, 130
Digital computers, 26, 28-29, 30, 59-60, 66, 85-86, 89, 91, 133
DIGFLY (P60) software, 54-5
DIGFLY2 software, 61
Digital Electronic Engine Controls (DEEC), 133
Direct Augmentation System (DIR), 96
Direct mode, 114
DO178B software standard, 129
Dominik, Daniel, 71
DOWNDIAG diagnostic program, 71
Draper, Charles Stark, 4, 12
Draper Laboratory, 18, 31, 33, 39, 46, 54; development and verification process, 36, 45; DIGFLY2, 61; EMP-004, 73; Erasable Memory Programs, 71; flight control software, 41, 48, 52-53, 96-97, 113; George Quinn, 99; Guidance and Control Division, 31; Iron Bird, 49; KSTART, 72; Phase II, 79, 87, 89; programming language, 90; RAVEN, 111; Shuttle Backup Flight System, 100, 107-108; software testing, 42, 44; see also Instrumentation Laboratory
Dryden, Hugh L., 125
Dryden Flight Research Center, 1, 22, 30, 33, 54, 75, 85, 97, 109; AP-101, 93, 105; design rules, 46; digital system, 29; F-8 Crusaders, 32, 48, 58; flight control, 52, 61, 66; Iron Bird, 47, 49; KECO, 59; Ken Szalai and, 50; KSTART, 70-71; landing training, 23; multi-computer system, 87, 103; Phase IB, 90-92; Phase II, 89, 100; Systems Analysis Branch, 110, and see also Flight Research Center
DSKY (Display and Keyboard Unit), 39, 40 ill., 47, 59, 61, 71-3
Dunham, Earl, 129
Dynamic Controls, Inc., 28
Dynamic stability, 4, 6

Eagle, see Lunar Module
Eckert, Presper, 26
Edwards Air Force Base, 21, 73 and passim
Egon blind bombing system, 14
Elevator Aileron Computers (ELAC), 128
Elliot, Jarrell, 116
Endeavour, 85 ill.
Enevoldson, Einar, 82, 113
Engle, Al, 54
ENIAC computer 26
Enterprise, 103 ill., 107, 111, 113
Envelope expansion, 105
Erasable Memory Programs, 71

F-4, 30, 33, 61-2, 64, 87-88, 88 ill., 98, 127, see also 680J
F-8, 33, 35-36, 44, 57, 66, 69, 70, 94, 100, 106, 112; Apollo and, 31; Supercritical Wing Project, 58, 75-6, 77 ill., 122 ill., 123; Backup Flight System, 61; Cooper-Harper rating of, 75; fly-by-wire system, 48, 51 ill., 59, 67, 76-77, 99, 127; Phase I, 45, 46, 65; Phase II, 85, 88-89; Phil Felleman and, 39; primary system, 63; software, 47, 53-54; test flights, 72, 74 58, 75-6, 123; F-8 DFBW team, 99 n. and passim, see also individual names
F-8C, frontispiece ill., vii ill., x ill., 55 ill., 57 ill., 61-2, 69 ill., 73, 77 ill., 103 ill., 122 ill., and see F-8; instrument panel, 74 ill.; refueling, 117 ill.

F-94, 22
F-14, 10, 58
F-15, 7
F-15 ACTIVE (Advanced Controls Technology for Integrated Vehicles), 133
F-16, 15, 38, 118, 133
F-16A, 64
F-16B, 64
F-16C, 117
F-16D, 117
F-18, 126, 134, and see F/A-18
F-100, 49
F-104, 32, 75, 106, 132
F-107, 61
F-111, 10, 23, 74
F-117A, 10, 15, 127 ill.
F/A-18 Systems Research Aircraft, 132-3
FAA (Federal Aviation Administration), 35, 46, 127, 129
Feedback control, 130
Felleman, Phil, 39, 46-8, 53, 71, 79, 90, 93
First Article Configuration Inspection (FACI), 42, 43
Fisher, Francis A., 100
Fleck, J.J., 38
Flight control system, 64, 65, 131
Flight Dynamics Laboratory, 26, 38
Flight Readiness Report, 75
Flight Research Center, 19, 21, 35, 57
Flight software, 48, 54
Fly-by-light, 133
FORTRAN program, 27, 90, 109, 119
FORTRAN-IV program, 116

Garman, John "Jack," 89, 93
Gemini Project, 17, 27, 30, 36, 89, 93
General Dynamics, 79, 82, 132
General Electric, 38, 44, 83, 87, 100
General Electric CP-32A, 89
Gera, Joseph, 116
Gibbs-Smith, Charles H., 6
Glenn, John H., Jr., 57
Glover, Richard, 97
Glycol system, 59
Gray, Richard E., 117
Ground Controlled Approaches (GCAs), 76, 81
Gulfstream jets, 23
Gyroscopic instruments, 24

Haber, Fritz, 15 n.
Haise, Fred W., Jr., 112, 114

HAL (computer language), 90, 93
Hall, Eldon, 39
Hancock, 57
Hand controller, 18
Hankins, James, 104
Hartmann, Gary L., 116
High Angle of Attack Research Vehicle (HARV), 133
Highly Integrated Digital Electronic Controls (HIDEC), 133
Highly Maneuverable Aircraft Technology (HiMAT), 127
Hoelzer, Helmut, 13, 15
Honeywell, Inc., 115-6; 601- 89; 801 "Alert," 71, 89
Hood, Raymond, 88
House Committee on Science and Astronautics, 83
Hudson, Fred, 94
Hunley, John "Dill," 133
Hydraulics Research and Manufacturing, 48, 66, 98, 100

IBM (International Business Machines), 27, 43, 91, 93-6, 105, 107, 111; IBM 360, 71
Icarus, 1
Inertial measurement unit, 19, 24, 54, 65, 67, 70
Inertial navigation system, 24
Inherent stability, 3-9, 8 ill.
Instrumentation Laboratory, 39, see also Draper Laboratory
Integrated circuit, 27
Intel 80486 computer, 128
Intercontinental ballistic missiles, 17
Intermetrics, 93
Iron Bird, 49, 55, 58 ill., 60, 70, 71, 73, 81, 87, 112 ill.; actuators, 52; AP-101s, 92 ill.; Krier, 101; McMurtry, 76; simulator, 47-48, 58; software of, 99, 119
Ishmael, Stephen D., 118-9, 134

J57 P20A engine, 67; P420 engine, 67
Jarvis, Calvin R., 22, 30, 31, 33, 35, 59, 88, 99, 134; backup flight system, 62, 63; control laws, 116; FAA – 129; Flight Research Center and, 46; flight software, 48, 60; KSTART, 71; LLRV program, 26; Phase I, 69, 73; Phase IB, 79, 87, 91; Phase II, 78, 83, 85-86, 101, 105-106; Shuttle program and, 89, 90; systems analysis, 49
JAS-39 Gripen, 23

Jenney, Gavin, 28
JetStar, 49-50, 88-9
JN-4 (Jenny), 9
Johnson Space Center, 23, 85, 100, 106, see also Manned Spacecraft Center

Kalman filters, 116
Kaufman, Dr. G., 116
KC-135 Stratotanker, 7
KECO, 59-60; cooling system, 86
Kennedy Space Center, 13
Kill Devil Hills, NC, 4
Kinematics, altitude, 119; rotational analysis, 119; translational, 119
Kitty Hawk, NC, 5
Korean conflict, 58
Krier, Gary, vii ill., 74; adaptive control law experiment, 116, 118; Aerospace Projects Office, 134; backup flight system, 62-63; Early Phase I, 75-78; F-4, 88; F-8, 73, 99, 101; House Committee on Science and Astronautics, 83; Iron Bird, 70; Phase II, 104, 105,106, 110; remote augmentation, 111; side stick, 80-81; Space Shuttle support, 108, 113, 114; YF-16 simulator, 79
KSTART, 70-73, 75, 77-8
Kurzhals, Peter R., 31, 87

Lancaster bomber, 15
Langley Research Center, 85, 87, 89-90, 100, 109, 115; Remotely Augmented Vehicle system, 108
Langley, Samuel Pierpont, 3
Larson, Richard, 97
Lateral instability, 6
Learjet, 88
Lee, Edmund, 130
Lenin, Vladimir, 130
Lewis Research Center, 83
Lightning Technologies, 83 ill.
Lillienthal, Otto, 3
Linear Variable Differential Transformers (LVDTs), 51, 66
Lock, Wilton P., 59, 62-64, 70, 73, 80, 97-98, 101, 120, 121, 134, 135
Longitudinal stability, 3, 11
Longitudinal instability, 8 ill.
LTV (formerly Ling Temco Vought), 48, 106
Luftwaffe, 13
Lunar Landing Research Vehicle (LLRV), 19, 21 ill., 21-2, 26, 29
Lunar Module, 17-9, 21

Lyapunov model, 115
Lyndon B. Johnson Space Center, see Johnson Space Center

M2-F1, 61
M2-F2, 61
Maine, Richard, 71
Majority organ, 37
Manke, John A., 113-4
Manly, Charles, 3
Manned Spacecraft Center (renamed Lyndon B. Johnson Space Center), 85, 89-90, 93
MARCO 4418 computer, 45
Marshall Space Flight Center, 44
Marx, Karl, 130
Mattingly, Kenneth, 82
Mauchly, John, 26
Maxim, Sir Hiram, 2-3, 5, 11
McDonnell-Douglas, 17
McMahon, William, 89
McMurtry, Thomas, 75-77, 79, 81, 106, 108, 110-11, 113, 134
McRuer, Duane, 76
Me 109, 6
Megna, Vincent, 47, 93, 108, 120, 121
Mekel, R., 115
Mercury, 17-8
Meteor jet, 24
Mid-value logic, 64
Miller, John, 93
Minuteman ICBM (Intercontinental Ballistic Missile), 27, 131
Mission Control, 54
Mistel, 15 n.
MIT (Massachusetts Institute of Technology), 39, 71, 79, 116; Electronic Systems Laboratory, 118; Instrumentation Laboratory, 18; see also Draper Laboratory
Mixing Computer, 15
Mode control panel, 74 ill.
Model following, 22
Montgomery, R.C., 115-16
Morris, James, 38, 62-3, 87
Motorola 68043 computer, 128

Nachmias, S., 115
NASA, 1, 25, 30, 38, 47, 132; Office of Advanced Research and Technology, 31; 802 (tail number of F-8 DFBW airplane), 55, 82; and Apollo, 40; computers, 61, 105; digital flight control, 66; F-8, 46, 89; F-104 Starfighters, 32; fly-by-wire project, 36;

Headquarters, 31, 33, 87; McDonnell-Douglas and, 17; Ken Szalai and, 50; MARCO 4418, 45; PGNCS, 18; software, 41-42, 44, 107; stage one simulator, 48; for individual Centers, see Center names (such as Johnson Space Center)
Navy, U.S.,123
New York University, 115
Newton-Raphson technique, 115
Northrop, 27
NOVA, 89

Oboe bombing system, 14
Obsitnik, Vincent, 91
Oestricher, Phillip, 82
Operation Paperclip, 15
Optimum Trajectory Research Experiment (OPTRE), 122

Parallel Channel Maximum Likelihood Estimation (PCMLE), 116
Parten, Richard, 90
Pascal computer language, 128
Peenemünde, 13
Petersen, Bill, 73,
Petersen, Kevin, 97, 110, 134
Peterson, Admiral Forrest S., 32
Peterson, Bruce, 69, 72-4, 77, 79, 81, 104
Phelps, James R., 55, 58-60, 67, 73, 75, 78, 80, 88, 91, 99, 134
Phugoid motion, 7
Pilcher, Percy, 3
Pilot Induced Oscillation (PIO), 75, 76, 85, 103, 112, 114
PL/M computer language, 128
Polaris, 18
Preliminary design review, 42
Primary Guidance Navigation and Control System (PGNCS), 18-9
Prince Charles, see Windsor
Program Change Notice, 53
Program Change Request, 53

Quinn, George, 99

RA-5C, 61
Rate gyros, 24
RAVADAPT, 116
Raytheon, 49, 60
RB-66, 88
RCA 215 computer, 89
Redstone missile, 17

Redundancy management, 103, 108, 126
Relaxed stability, 7
Reliability in computers, 36-39, 45-46, see also AP-101
Remote augmentation, 111-12
Remotely Augmented Vehicle, 85-6, 96, 108-110, 127
Remotely Augmented Vehicle Experimental Norm (RAVEN), 110-11
Remotely Piloted Vehicle Facility, 122
Rensselaer Polytechnic Institute, 116
Resident Backup Software (REBUS), 85, 114, 120-22
Response feedback, 22
Richardson, Bruce, 73
Rockwell International Corp., 100, 106-7
Roll Rate Command (RRC), 78, 81
Royal Aeronautical Society, 12
Royal Aircraft Establishment, 111
Rushby, John, 35-6

S-3A, 88
Sampling theory, 26
Satterfield, James, 93
Saturn V, 38, 44-5
Sayers, W.H., 5
Schneider, Edward, 103, 121-23, 134
Scott, David, 101
Secondary actuators, 97-8
Sensor analytic redundancy management, 85, 114, 118-120
SFENA, 128
Shannon, Claude, 130
Shea, Joseph, 44
SHERLOCK diagnostic program, 71-2
Shuttle, 83, 85 ill., 87, 89, 103 ill., 108; Apollo Lunar Module and, 17; Approach and Landing Tests, 106, 111-14; Backup Flight System, 100, 107; digital redundancy, 85; IBM-Houston, 43; orbiter, 23, 107; side stick, 65; software, 93, 96
Shuttle Carrier Aircraft, 103 ill.
Side stick, 69, 79-82, 80 ill., 86, 109; side-stick controllers, 11
Singer-Kearfott SKC-2000 computer, 90-1
Skylab, 61
Smith, Rogers, 121, 134
Sneak circuit analysis, 100-1
Software Engineering Institute, see Carnegie
Software Verification Reports, 97
Soviet Union, 107
Space Transportation System, 93

Spad, 5
Sperry, 11, 23, 61-2, 64, 67, 70, 89, 98, 100-1; 1819A computer, 89; analog system, 99
Spoiler Elevator Computers, 128
Sputnik, 16
SR-71, 46
Stability augmentation, 69
Stability Augmentation System (SAS), 10, 24, 61-2, 72-6, 81, 96, 106, 114, 120
Stalin, Joseph, 130
Stein, Gunter, 116
Stick shaping, 51, 73
Structural mode control, 23
Sullivan, Frank J., 31
Supercritical Wing Project, 58, 75-6, 123
Supersonic fighter, 57
Supplementary Type Certificate, 88
Surveyor, 18
Survivable Flight Control System, 30
Szalai, Kenneth J., 22, 46, 50-53, 66, 73, 78, 95, 105, 116, 133, 134; active flight control, 126; AP-101s, 94; and CF-105 Arrow, 63; digital computer, 30, 60, 89, 91; digital redundancy, 86; DOWNDIAG, 71; F-8, 32-33; Flight Readiness Report (1972), 75; Phase I, 49; Phase IB, 87; Phase II, 101; REBUS, 120; Shuttle software, 107, 108; Software Verification Reports, 97

T-33, 22
T-38, 79
T-39, 88
Technical University of Darmstadt, 13
Teledyne 43M computer, 91
Thompson, Milton O., 104
Thomson-CSF, 128
Three-axis flight control, 16
Tindall, Howard W. "Bill," 41, 89
Titan booster, 17
Translational dynamics, 119
Transonic region, 7
Transport delays, 112-14
TRW, Inc., 42, 44-5

U.S.S. Hancock, 100
U.S.S Ticonderoga, 57
University of Pennsylvania, 26
Unstable vehicle, 12 ill.
UPSUM (checksum verification), 72

Vandling, Gib, 91
Variable reliability, 49

Variable stability, 22
Varian, 116; V-73 minicomputer, 110
Versailles Treaty, 13
Vincenti, Walter A., 131
von Braun, Wernher, 13, 15
von Neumann, John, 37-9
Voyagers, 38

Weber, Max, 130
Western Society of Engineers, 1
When Worlds Collide (film), 25
Wiener, Norbert, 130
Willsky, A., 118
Wilson, Ronald "Joe," 97, 104, 134-5
Windsor, Charles Philip Arthur George, Prince of Wales, 112
Wing warping, 4
Workshop on NASA Advanced Flight Control Systems Experience, 126
World War I, 5, 57
World War II, 6, 13, 15, 26, 58, 130
Wright brothers, 1, 3-4, 6, 12-3
Wright Flyer, 1, 2 ill., 4
Wright lecture, 12
Wright, Orville, 3-5
Wright, Wilbur, 1, 5
Wright-Patterson Air Force Base, 28, 38, 62

X-15, 17, 21, 31-2
X-20 Dyna-Soar, 17, 64
X-24A, 61
X-29, 123
X-31, 33
XB-47 Stratojet, 24
XB-49, 24
XDS 9300 computer, 47

Yaw damper, 16
YC-14, 95 ill., 134
YF-16, 65, 69, 79-83
YF-17, 82

Zimmerman, William, 93
Zola, Edward, 91

THE NASA HISTORY SERIES

Reference Works, NASA SP-4000:

Grimwood, James M. *Project Mercury: A Chronology.* (NASA SP-4001, 1963).

Grimwood, James M., and Hacker, Barton C., with Vorzimmer, Peter J. *Project Gemini Technology and Operations: A Chronology.* (NASA SP-4002, 1969).

Link, Mae Mills. *Space Medicine in Project Mercury.* (NASA SP-4003, 1965).

Astronautics and Aeronautics, 1963: Chronology of Science, Technology, and Policy. (NASA SP-4004, 1964).

Astronautics and Aeronautics, 1964: Chronology of Science, Technology, and Policy. (NASA SP-4005, 1965).

Astronautics and Aeronautics, 1965: Chronology of Science, Technology, and Policy. (NASA SP-4006, 1966).

Astronautics and Aeronautics, 1966: Chronology of Science, Technology, and Policy. (NASA SP-4007, 1967).

Astronautics and Aeronautics, 1967: Chronology of Science, Technology, and Policy. (NASA SP-4008, 1968).

Ertel, Ivan D., and Morse, Mary Louise. *The Apollo Spacecraft: A Chronology, Volume I, Through November 7, 1962.* (NASA SP-4009, 1969).

Morse, Mary Louise, and Bays, Jean Kernahan. *The Apollo Spacecraft: A Chronology, Volume II, November 8, 1962-September 30, 1964.* (NASA SP-4009, 1973).

Brooks, Courtney G., and Ertel, Ivan D. *The Apollo Spacecraft: A Chronology, Volume III, October 1, 1964-January 20, 1966.* (NASA SP-4009, 1973).

Ertel, Ivan D., and Newkirk, Roland W., with Brooks, Courtney G. *The Apollo Spacecraft: A Chronology, Volume IV, January 21, 1966-July 13, 1974.* (NASA SP-4009, 1978).

Astronautics and Aeronautics, 1968: Chronology of Science, Technology, and Policy. (NASA SP-4010, 1969).

Newkirk, Roland W., and Ertel, Ivan D., with Brooks, Courtney G. *Skylab: A Chronology.* (NASA SP-4011, 1977).

Van Nimmen, Jane, and Bruno, Leonard C., with Rosholt, Robert L. *NASA Historical Data Book, Volume I: NASA Resources, 1958-1968.* (NASA SP-4012, 1976, rep.ed. 1988).

Ezell, Linda Neuman. *NASA Historical Data Book, Volume II: Programs and Projects, 1958-1968.* (NASA SP-4012, 1988).

Ezell, Linda Neuman. *NASA Historical Data Book, Volume III: Programs and Projects, 1969-1978.* (NASA SP-4012, 1988).

Gawdiak, Ihor Y., with Fedor, Helen. Compilers. *NASA Historical Data Book, Volume IV: NASA Resources, 1969-1978.* (NASA SP-4012, 1994).

Astronautics and Aeronautics, 1969: Chronology of Science, Technology, and Policy. (NASA SP-4014, 1970).

Astronautics and Aeronautics, 1970: Chronology of Science, Technology, and Policy. (NASA SP-4015, 1972).

Astronautics and Aeronautics, 1971: Chronology of Science, Technology, and Policy. (NASA SP-4016, 1972).

Astronautics and Aeronautics, 1972: Chronology of Science, Technology, and Policy. (NASA SP-4017, 1974).

Astronautics and Aeronautics, 1973: Chronology of Science, Technology, and Policy. (NASA SP-4018, 1975).

Astronautics and Aeronautics, 1974: Chronology of Science, Technology, and Policy. (NASA SP-4019, 1977).

Astronautics and Aeronautics, 1975: Chronology of Science, Technology, and Policy. (NASA SP-4020, 1979).

Astronautics and Aeronautics, 1976: Chronology of Science, Technology, and Policy. (NASA SP-4021, 1984).

Astronautics and Aeronautics, 1977: Chronology of Science, Technology, and Policy. (NASA SP-4022, 1986).

Astronautics and Aeronautics, 1978: Chronology of Science, Technology, and Policy. (NASA SP-4023, 1986).

Astronautics and Aeronautics, 1979-1984: Chronology of Science, Technology, and Policy. (NASA SP-4024, 1988).

Astronautics and Aeronautics, 1985: Chronology of Science, Technology, and Policy. (NASA SP-4025, 1990).

Noordung, Hermann. *The Problem of Space Travel: The Rocket Motor.* Stuhlinger, Ernst, and Hunley, J.D., with Garland, Jennifer. Editor. (NASA SP-4026, 1995).

Astronautics and Aeronautics, 1986-1990: A Chronology. (NASA SP-4027, 1997).

Rumerman, Judy A. Compiler. *NASA Historical Data Book, 1979-1988: Volume V, NASA Launch Systems, Space Transportation, Human Spaceflight, and Space Science.* (NASA SP-4012, 1999).

Management Histories, NASA SP-4100:

Rosholt, Robert L. *An Administrative History of NASA, 1958-1963.* (NASA SP-4101, 1966).

Levine, Arnold S. *Managing NASA in the Apollo Era.* (NASA SP-4102, 1982).

Roland, Alex. *Model Research: The National Advisory Committee for Aeronautics, 1915-1958.* (NASA SP-4103, 1985).

Fries, Sylvia D. *NASA Engineers and the Age of Apollo.* (NASA SP-4104, 1992).

Glennan, T. Keith. *The Birth of NASA: The Diary of T. Keith Glennan.* Hunley, J.D. Editor. (NASA SP-4105, 1993).

Seamans, Robert C., Jr. *Aiming at Targets: The Autobiography of Robert C. Seamans, Jr.* (NASA SP-4106, 1996)

Project Histories, NASA SP-4200:

Swenson, Loyd S., Jr., Grimwood, James M., and Alexander, Charles C. *This New Ocean: A History of Project Mercury.* (NASA SP-4201, 1966).

Green, Constance McL., and Lomask, Milton. *Vanguard: A History.* (NASA SP-4202, 1970; rep. ed. Smithsonian Institution Press, 1971).

Hacker, Barton C., and Grimwood, James M. *On Shoulders of Titans: A History of Project Gemini.* (NASA SP-4203, 1977).

Benson, Charles D. and Faherty, William Barnaby. *Moonport: A History of Apollo Launch Facilities and Operations.* (NASA SP-4204, 1978).

Brooks, Courtney G., Grimwood, James M., and Swenson, Loyd S., Jr. *Chariots for Apollo: A History of Manned Lunar Spacecraft.* (NASA SP-4205, 1979).

Bilstein, Roger E. *Stages to Saturn: A Technological History of the Apollo/Saturn Launch Vehicles.* (NASA SP-4206, 1980).

SP-4207 not published.

Compton, W. David, and Benson, Charles D. *Living and Working in Space: A History of Skylab.* (NASA SP-4208, 1983).

Ezell, Edward Clinton, and Ezell, Linda Neuman. *The Partnership: A History of the Apollo-Soyuz Test Project.* (NASA SP-4209, 1978).

Hall, R. Cargill. *Lunar Impact: A History of Project Ranger.* (NASA SP-4210, 1977).

Newell, Homer E. *Beyond the Atmosphere: Early Years of Space Science.* (NASA SP-4211, 1980).

Ezell, Edward Clinton, and Ezell, Linda Neuman. *On Mars: Exploration of the Red Planet, 1958-1978.* (NASA SP-4212, 1984).

Pitts, John A. *The Human Factor: Biomedicine in the Manned Space Program to 1980.* (NASA SP-4213, 1985).

Compton, W. David. *Where No Man Has Gone Before: A History of Apollo Lunar Exploration Missions.* (NASA SP-4214, 1989).

Naugle, John E. *First Among Equals: The Selection of NASA Space Science Experiments.* (NASA SP-4215, 1991).

Wallace, Lane E. *Airborne Trailblazer: Two Decades with NASA Langley's Boeing 737 Flying Laboratory.* (NASA SP-4216, 1994).

Butrica, Andrew J. Editor. *Beyond the Ionosphere: Fifty Years of Satellite Communication*. (NASA SP-4217, 1997).

Butrica, Andrews J. *To See the Unseen: A History of Planetary Radar Astronomy.* (NASA SP-4218, 1996).

Mack, Pamela E. Editor. *From Engineering Science to Big Science: The NACA and NASA Collier Trophy Research Project Winners.* (NASA SP-4219, 1998).

Reed, R. Dale. With Lister, Darlene. *Wingless Flight: The Lifting Body Story.* (NASA SP-4220, 1997).

Heppenheimer, T.A. *The Space Shuttle Decision: NASA's Search for a Reusable Space Vehicle.* (NASA SP-4221, 1999).

Hunley, J.D. Editor. *Toward Mach 2: The Douglas D-558 Program.* (NASA SP-4222, 1999).

Swanson, Glen E. Editor. *"Before this Decade is Out...": Personal Reflections on the Apollo Program* (NASA SP-4223, 1999).

Center Histories, NASA SP-4300:

Rosenthal, Alfred. *Venture into Space: Early Years of Goddard Space Flight Center.* (NASA SP-4301, 1985).

Hartman, Edwin, P. *Adventures in Research: A History of Ames Research Center, 1940-1965.* (NASA SP-4302, 1970).

Hallion, Richard P. *On the Frontier: Flight Research at Dryden, 1946-1981.* (NASA SP-4303, 1984).

Muenger, Elizabeth A. *Searching the Horizon: A History of Ames Research Center, 1940-1976.* (NASA SP-4304, 1985).

Hansen, James R. *Engineer in Charge: A History of the Langley Aeronautical Laboratory, 1917-1958.* (NASA SP-4305, 1987).

Dawson, Virginia P. *Engines and Innovation: Lewis Laboratory and American Propulsion Technology.* (NASA SP-4306, 1991).

Dethloff, Henry C. *"Suddenly Tomorrow Came...": A History of the Johnson Space Center.* (NASA SP-4307, 1993).

Hansen, James R. *Spaceflight Revolution: NASA Langley Research Center from Sputnik to Apollo.* (NASA SP-4308, 1995).

Wallace, Lane E. *Flights of Discovery: 50 Years at the NASA Dryden Flight Research Center.* (NASA SP-4309, 1996).

Herring, Mack R. *Way Station to Space: A History of the John C. Stennis Space Center.* (NASA SP-4310, 1997).

Wallace, Harold D., Jr. *Wallops Station and the Creation of the American Space Program.* (NASA SP-4311, 1997).

Wallace, Lane E. *Dreams, Hopes, Realities: NASA's Goddard Space Flight Center, The First Forty Years.* (NASA SP-4312).

General Histories, NASA SP-4400:

Corliss, William R. *NASA Sounding Rockets, 1958-1968: A Historical Summary.* (NASA SP-4401, 1971).

Wells, Helen T., Whiteley, Susan H., and Karegeannes, Carrie. *Origins of NASA Names.* (NASA SP-4402, 1976).

Anderson, Frank W., Jr. *Orders of Magnitude: A History of NACA and NASA, 1915-1980.* (NASA SP-4403, 1981).

Sloop, John L. *Liquid Hydrogen as a Propulsion Fuel, 1945-1959.* (NASA SP-4404, 1978).

Roland, Alex. *A Spacefaring People: Perspectives on Early Spaceflight.* (NASA SP-4405, 1985).

Bilstein, Roger E. *Orders of Magnitude: A History of the NACA and NASA, 1915-1990.* (NASA SP-4406, 1989).

Logsdon, John M. Editor. With Lear, Linda J., Warren-Findley, Jannelle, Williamson, Ray A., and Day, Dwayne A. *Exploring the Unknown: Selected Documents in the History of the U.S. Civil Space Program, Volume I, Organizing for Exploration.* (NASA SP-4407, 1995).

Logsdon, John M. Editor. With Day, Dwayne A., and Launius, Roger D. *Exploring the Unknown: Selected Documents in the History of the U.S. Civil Space Program, Volume II, Relations with Other Organizations.* (NASA SP-4407, 1996).

Logsdon, John M. Editor. With Launius, Roger D., Onkst, David H., and Garber, Stephen. *Exploring the Unknown: Selected Documents in the History of the U.S. Civil Space Program, Volume III, Using Space.* (NASA SP-4407, 1998).

www.ingramcontent.com/pod-product-compliance
Lightning Source LLC
Chambersburg PA
CBHW081236180526
45171CB00005B/443